LABORATORY MANUAL

Charles H. Corwin
American River College

Introductory Chemistry

CONCEPTS AND CRITICAL THINKING

SIXTH EDITION

CHARLES H. CORWIN

PEARSON

Boston Columbus Indianapolis New York San Francisco Upper Saddle River
Amsterdam Cape Town Dubai London Madrid Milan Munich Paris Montréal Toronto
Delhi Mexico City São Paulo Sydney Hong Kong Seoul Singapore Taipei Tokyo

Editor in Chief: Adam Jaworski
Executive Editor: Jeanne Zalesky
Senior Project Editor: Jennifer Hart
Senior Marketing Manager: Jonathan Cottrell
Editorial Assistant: Fran Falk
Marketing Assistant: Nicola Houston
Managing Editor, Chemistry and Geosciences: Gina M. Cheselka
Project Manager, Production: Maureen Pancza
Cover Designer: Seventeenth Street Studios
Operations Specialist: Jeffrey Sargent
Cover: Pyramids of white salt, Salar de Uyuni in southwest Bolivia; Kazuyoshi Nomachi/Corbis

4 5 6 7 8 9 10—V064— 18 17 16 15 14

PEARSON

www.pearsonhighered.com

ISBN-10: 0-321-75094-2; ISBN-13: 978-0-321-75094-5

LABORATORY MANUAL

Introductory Chemistry

CONCEPTS AND CRITICAL THINKING

SIXTH EDITION

EXPERIMENT

Contents

* Assigned Unknown

APPENDICES

EXPERIMENT

Preface

At a chemistry conference, an instructor using the lab manual mentioned that the experiments were remarkably "bullet-proof." I responded that our department instructs over 1000 intro chem students in the laboratory each year, and our chemistry program employs rotating adjunct faculty who bring a fresh set of eyes to the experiments we supervise. This constant turnover affords ongoing feedback and the opportunity to further fine tune each procedure and assignment.

The ***Pearson Laboratory Manual for Introductory Chemistry, 6/E,*** continues to evolve with increased sensitivity to environmental and safety concerns in the laboratory. In this edition, we have introduced "green chemicals" and a recycle icon appears in the margin of each procedure as a reminder to students that chemicals are to be disposed of in the waste containers provided.

What's New in This Edition?

Responding to environmental regulations and instructor reviews of the previous edition, students will benefit from a variety of new content in the Sixth Edition including:

- New ***Environmental Icons*** to alert students to recycle chemical waste.
- New ***Instructor Demonstrations*** for procedures that reduce chemical waste.
- New ***Experimental Procedures*** that simplify tasks and provide a better work flow.
- New ***Prelaboratory Assignments*** to help students prepare for an experiment.
- New ***Postlaboratory Assignments*** to synthesize the principles in an experiment.
- New ***Experiment 25***, a comprehensive review of lab techniques as a practical final exam.
- New ***Appendix H***, directions for keeping a laboratory notebook.

RECYCLE
Chemical
Waste

What Features Are in Each Experiment?

To help introductory chemistry students to be organized in laboratory and to have a safe experience, each experiment has the following features:

- A set of ***Objectives*** to help students focus on experimental activities.
- A ***Discussion*** with example exercises to help students with calculations and equations.
- A list of ***Equipment and Chemicals*** to organize the experimental materials.
- A stepwise ***Procedure*** to systematically guide the flow of activity.
- A ***Prelaboratory Assignment*** with safety precautions to prepare students before lab.
- A ***Data Table*** to help students learn to accurately record observations and measurements.
- A ***Postlaboratory Assignment*** to reinforce the principles in the experiment.

Instructor's Manual and Quiz Item File

A complementary Instructor's Manual is provided with each adoption of the laboratory manual. The ***Annotated Instructor's Manual*** contains the following for each experiment: suggested unknowns and directions for dispensing and preparing solutions, sample data tables, answers to postlaboratory assignments, and a *Quiz Item File* containing over 500 class-tested questions.

The ***Annotated Instructor's Manual*** also contains a *Master List of Reagents & Suppliers* for all chemicals required for each experiment, along with directions for the preparation of aqueous solutions. A list of websites, addresses, and phone numbers of suppliers for chemicals and equipment is provided to assist stockroom personnel.

Acknowledgments

These latest experiments reflect the suggestions of instructors and students who have e-mailed comments. In addition, I am fortunate to have the shared expertise of colleagues including: Kristin Casale, Darren Gottke, Ronald Grider, Tamilyn Hong, Greg Jorgensen, Michael Maddox, Dianne Meador, Chris Meadows, Edmund Niedzinski, Michael Payne, Karen Pesis, Deboleena Roy, Daniel Stewart, Brian Weissbart, Veronica Wheaton, and Linda Zarzana.

A successful laboratory program is helped immeasurably by capable stockroom personnel. The ongoing refinement of these experiments has been facilitated by our stockroom lab technicians, Cuong Bui, Chris Douglas, and Ed Hege. I, along with my colleagues, greatly appreciate the support provided by Cuong, Chris, Ed, and staff.

I am grateful to Jennifer Hart, Pearson senior project editor, who kept everything flowing seamlessly from initial reviews of the previous edition through the production of this latest edition. And a special thanks to Dr. Kent McCorkle, Fresno City College, who did an accuracy check on the entire final manuscript.

REVIEWERS FOR THE SIXTH EDITION

Elaine M. Alfonsetti
Broome Community College

Edward L. Barnes
Fayetteville Technical Community College

Melinda Neal
Cowley College

Raymond Sadeghi
University of Texas, San Antonio

David Baker
Delta College

Melekeh Nasiri
University of California, Davis

Edmund J. Niedzinski
American River College

Clarissa Sorensen-Unruh
Central New Mexico College

Charles H. Corwin
Department of Chemistry
American River College
Sacramento, CA 95841
corwinc@arc.losrios.edu

LABORATORY MANUAL

Introductory Chemistry

CONCEPTS AND CRITICAL THINKING

SIXTH EDITION

EXPERIMENT

Safety Precautions

With proper precautions, a chemistry laboratory should not be a dangerous place. If you do the prelaboratory assignment and check your answers in Appendix J before coming to lab, the laboratory should be safe to do your experiments. The following rules are common sense.

1. Wear approved safety goggles while working in the laboratory.
2. Wear shoes (not sandals) while working in the laboratory.
3. Do not bring food or drink into the laboratory.
4. Locate the fire extinguisher(s).
5. Locate the first-aid equipment.
6. Do not perform unauthorized experiments.
7. Do not smell a gas directly; instead gently waft the vapor toward your nose.

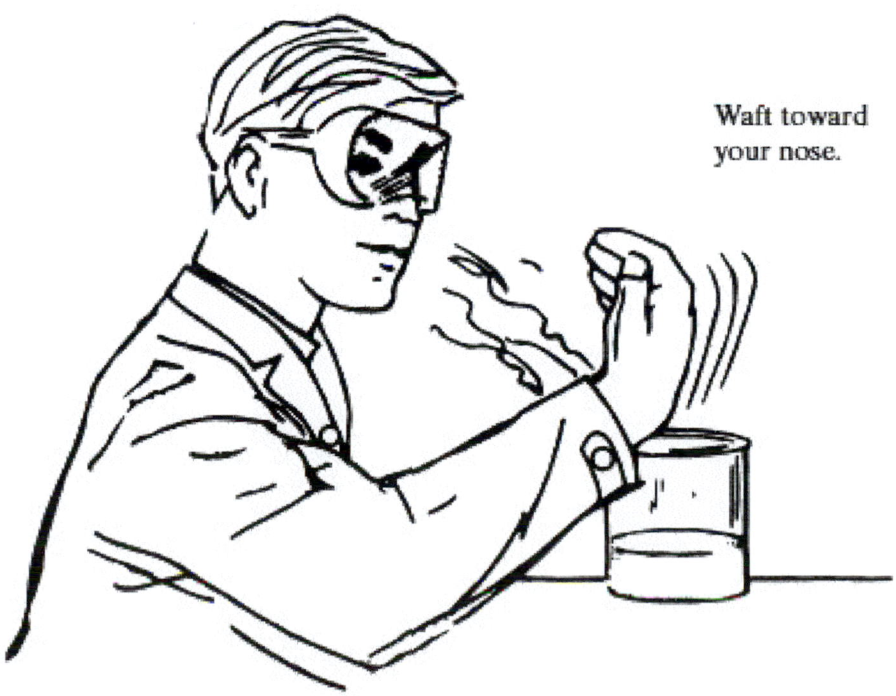

8. Perform experiments that produce a gas under a fume hood.
9. When heating a test tube, point the open end in a safe direction.

10. Always pour an acid into water—not water into acid.
11. Clean up broken glass immediately.
12. Do not use an organic liquid near an open flame in the laboratory. Organic liquids, such as acetone and alcohol, are highly flammable.
13. If you contact a chemical, wash immediately with water and notify the Instructor.
14. Notify the Instructor immediately in the event of an accident.

EXPERIMENT

[redacted]

Locker Inventory

EQUIPMENT	QUANTITY
beakers, 100, 150, 250, 400, 600, 1000 mL	1 each
clay triangle	1
crucible and cover	1
crucible tongs	1
dropper pipet	1
Erlenmeyer flasks, 125 mL	3
Erlenmeyer flasks, 250 mL	3
evaporating dish	1
Florence flask, 1000 mL	1
long-stem funnel	1
small plastic buret funnel (optional)	1
graduated cylinder, 100 mL	1
litmus paper	1
stirring rod, thin glass	1
stirring rod with rubber policeman	1
test tubes, 16 x 150 mm	2
test tubes, 13 x 100 mm	6
test tube brush	1
test tube holder	1
test tube rack	1
thermometer, 110°C	1
wash bottle, plastic	1
watchglass, ~150 mm	1
wire gauze	1

ASSIGNED LOCKER # ________________ NAME ____________________________

COMBINATION ______________________ SECTION ____________________________

COMMON LABORATORY EQUIPMENT

beaker

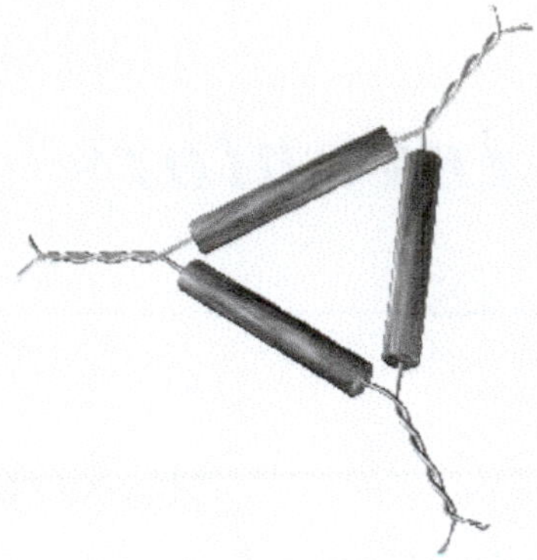

clay triangle

crucible and cover

dropper pipet

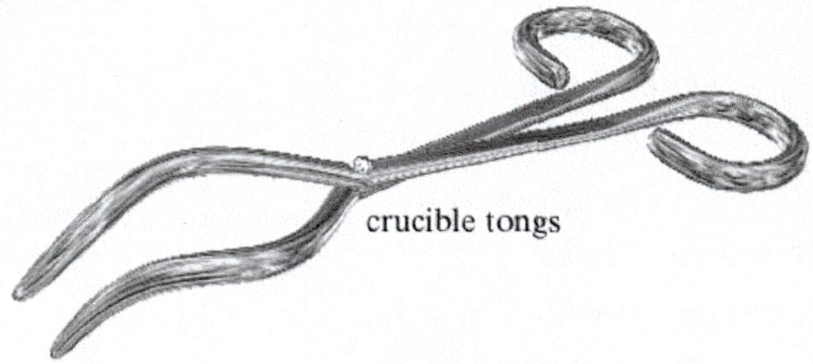

crucible tongs

Erlenmeyer flask

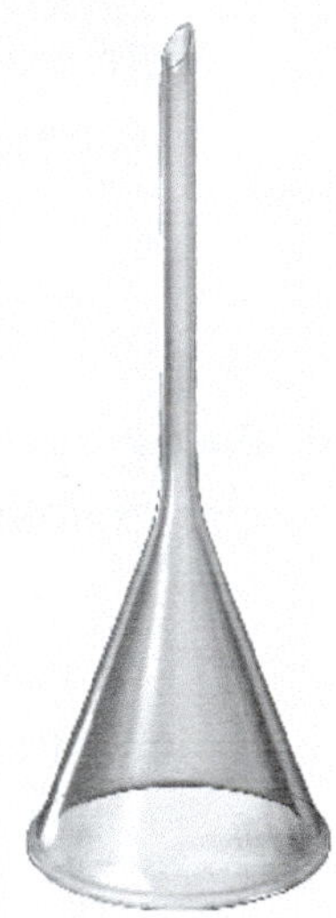

long-stem funnel

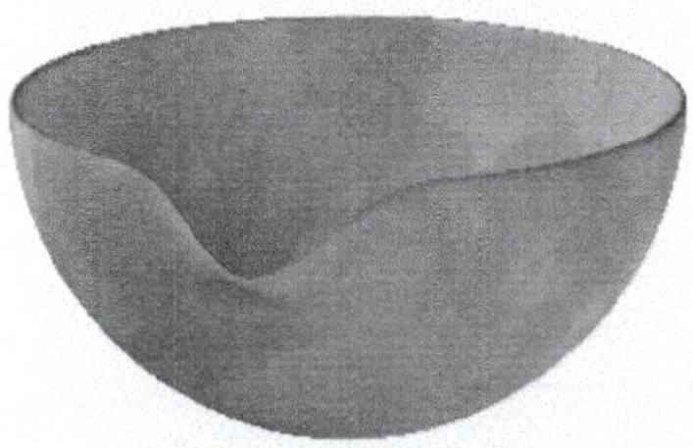

evaporating dish

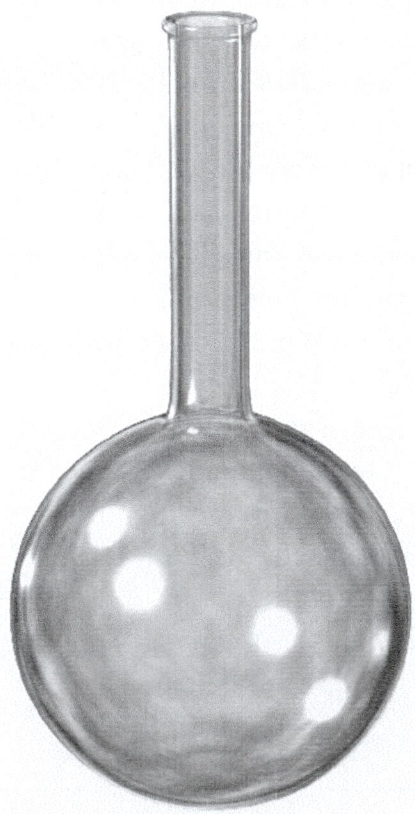

Florence flask

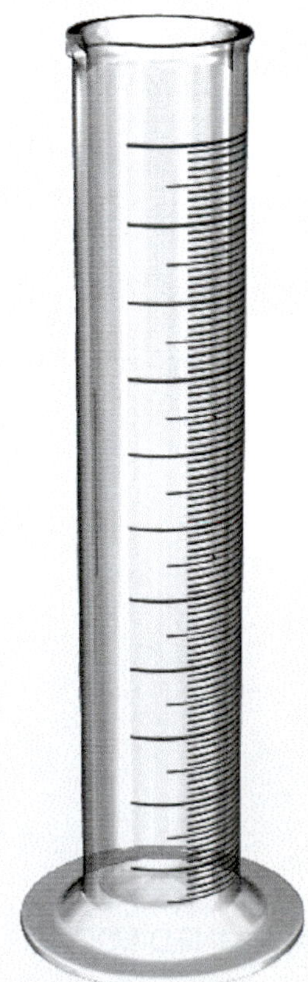

graduated cylinder

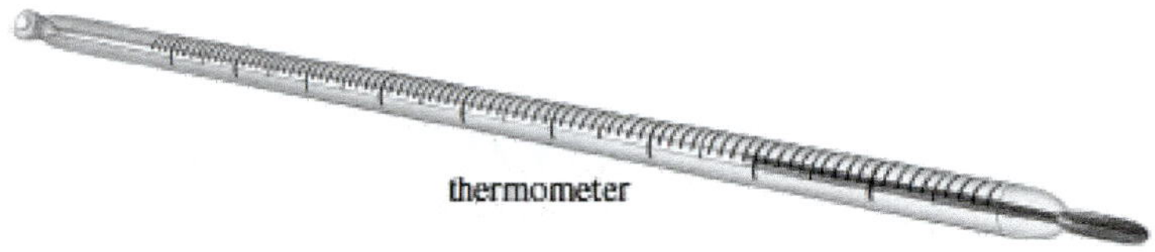
thermometer

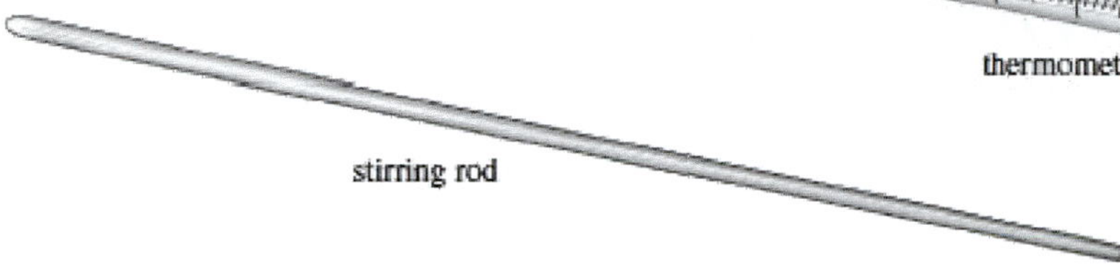
stirring rod

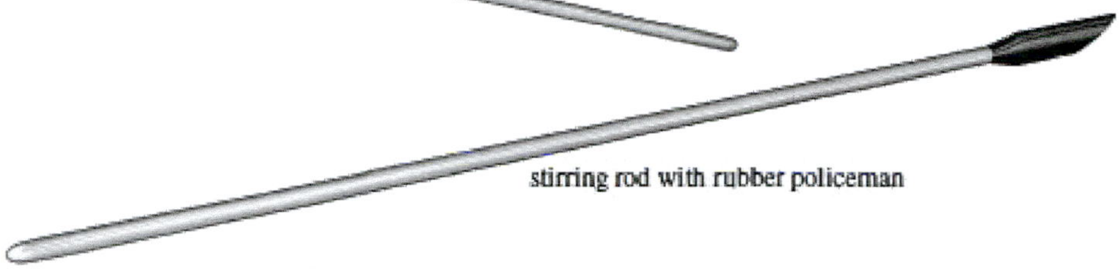
stirring rod with rubber policeman

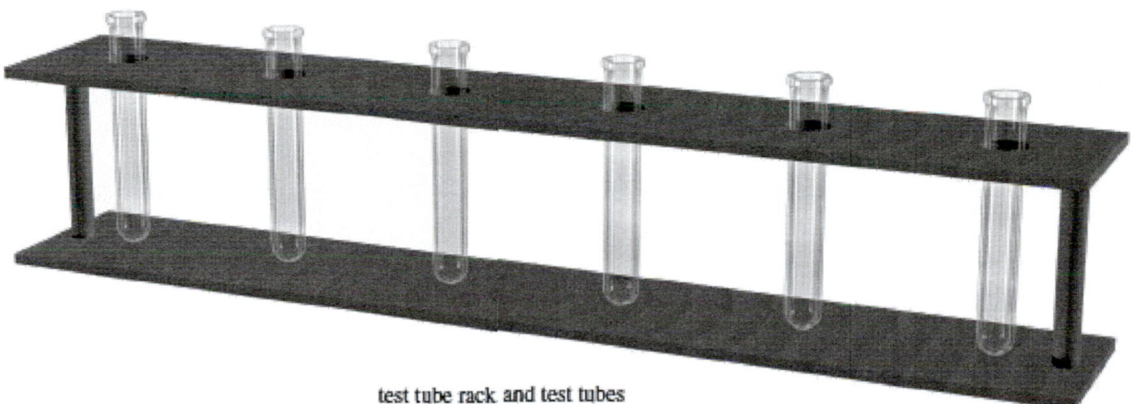
test tube rack and test tubes

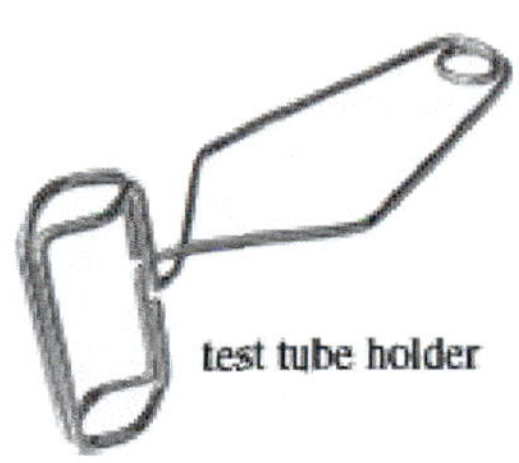
test tube holder

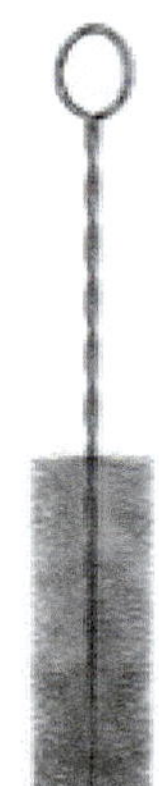
test tube brush

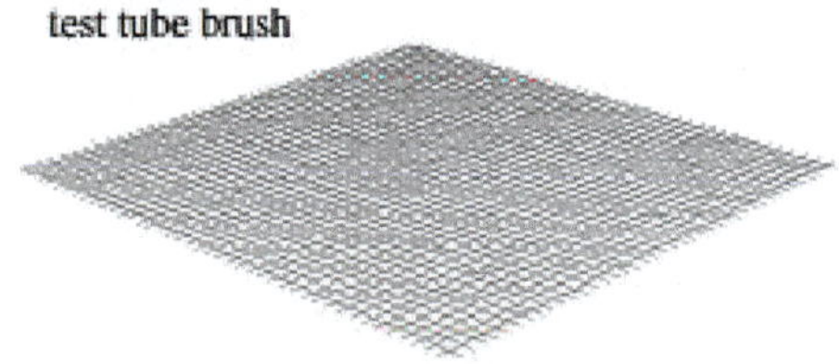

wash bottle

watchglass

wire gauze

EXPERIMENT
[redacted]

Waste Disposal

INORGANIC CHEMICAL WASTE

RECYCLE
Chemical
Waste

Inorganic chemical waste includes solid compounds *not* containing carbon, aqueous solutions, acids, and bases. Each of these chemicals should be disposed of in an "**inorganic waste**" container.

ORGANIC CHEMICAL WASTE

Organic chemical waste includes solid compounds containing carbon, and organic solvents such as acetone, alcohol, heptane, and hexane. Each of these chemicals should be disposed of in an "**organic waste**" container.

BROKEN GLASS

As mentioned in the *Safety Precautions*, broken glass should be cleaned up immediately. The laboratory has a broom and dustpan for sweeping up small shards of glass.

Glassware with cracks or sharp edges should not be used. Dispose of broken or cracked glassware in the "**broken glass**" container, and ***not*** in the paper trash container.

EXPERIMENT

1

Introduction to Chemistry

OBJECTIVES

- To gain experience in recording data and explaining observations.
- To develop skill in handling glassware and transferring chemicals.
- To become familiar with basic safety precautions in the laboratory.

DISCUSSION

Chemistry is a science that studies the composition and properties of matter. We can define the term **science** as the methodical exploration of nature and the logical interpretation of the observations. In an **experiment**, scientists gather data and carefully record observations under controlled conditions. After an experiment, scientists formulate a tentative proposal, or **hypothesis**, to explain the data. If additional experiments support the original proposal, a hypothesis may be elevated to a scientific principle, or **theory**. This stepwise procedure is known as the **scientific method** and can be summarized as follows:

Step 1: Perform a planned experiment, make observations, and record data.
Step 2: Analyze the data, and propose a tentative hypothesis to explain the observations.
Step 3: Conduct additional experiments to test the hypothesis. If the evidence supports the initial proposal, the hypothesis may become a theory.

We should note that scientists exercise caution before accepting a theory. Experience has shown that nature reveals her secrets slowly and only after considerable probing. The following example exercise illustrates the scientific method.

Example Exercise 1.1 • The Scientific Method

A blue turquoise crystal is heated in a test tube. A colorless, odorless liquid collects inside the test tube and the blue crystal turns into a white powder.

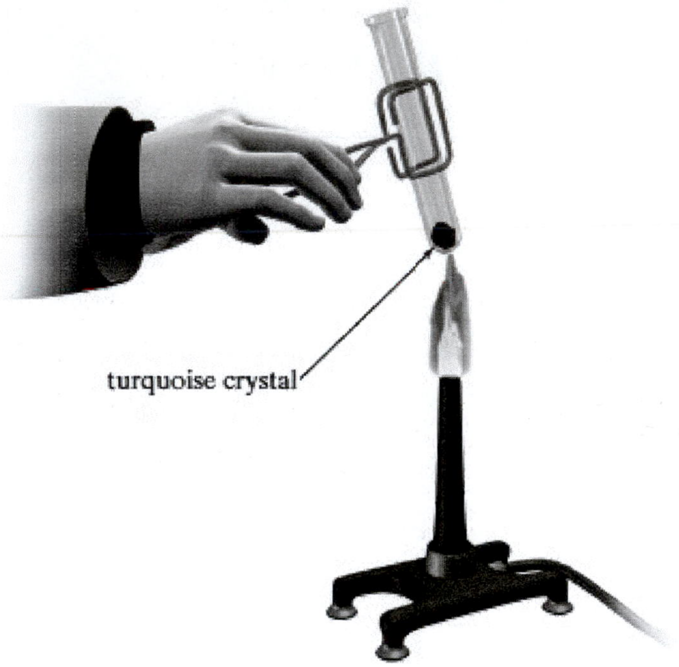

Figure 1.1 Heating Blue Turquoise The blue turquoise crystal changes to a white powder and releases a colorless, odorless liquid after heating.

Observation

- A colorless, odorless liquid is observed after heating the blue crystal.
- The blue crystal changes to a white powder after heating.

Hypothesis

- Heating a turquoise crystal produces water.
- The blue crystal changes to a white powder by losing water.

EQUIPMENT and CHEMICALS

A. Instructor Demonstrations

- tall glass cylinder
- large Erlenmeyer flask + stopper
- glass stirring rod
- 150-mL beaker
- matches
- fire extinguisher
- mortar and pestle
- wash bottle
- evaporating dish
- cupric sulfate solution, 0.1 *M* $CuSO_4$
- ammonium hydroxide, 6 *M* NH_4OH
- copper penny (pre-1982 mint date)
- conc. nitric acid, 16 *M* HNO_3
- sugar, powdered $C_{12}H_{22}O_{11}$
- conc. sulfuric acid, 18 *M* H_2SO_4
- ethyl alcohol, CH_3CH_2OH
- ammonium nitrate, solid NH_4NO_3
- zinc, Zn powder

B. Student Experiments

- 1000-mL Florence flask + stopper and disappearing blue solution
- 13 x 100 mm test tubes (2)
- spatulas (2)
- 250-mL beaker
- 100-mL beaker
- ball-and-stick models
- (10 g glucose in 300 mL 0.5 *M* KOH + 10 mL of 0.1 g/L methylene blue solution)
- ammonium chloride, solid NH_4Cl
- calcium chloride, solid $CaCl_2$
- iron, Fe nail
- calcium, Ca metal
- copper(II) sulfate solution, 0.1 *M* $CuSO_4$

PROCEDURE

A. Instructor Demonstrations

The following chemical demonstrations should be performed by the Instructor. Students are to record observations and propose a hypothesis to explain their observation.

1. *Cold Heat.* Add 40 mL of ethyl alcohol to 60 mL of water in a 150-mL beaker. Soak a cotton handkerchief in the alcohol solution and squeeze out the excess. Hold the handkerchief with crucible tongs, dim the room lights, and ignite.

   Note: The Instructor may wish to point out the location of the fire extinguisher and flammable solvents in the laboratory. Students should try to explain why the cotton handkerchief, soaked in alcohol, does not burn.

RECYCLE **Chemical Waste**

2. *Black Foam.* Half fill a 150-mL beaker with household powdered sugar. Add 15 mL of concentrated sulfuric acid and stir slowly with a glass rod.

   Note: Students should try to identify the black foam. The formula for ordinary powdered sugar is $C_{12}H_{22}O_{11}$.

3. *Copper Smog.* Drop a copper penny (pre-1982 mint date) into a large Erlenmeyer flask. Pour a few milliliters of concentrated nitric acid into the flask to cover the penny and insert a rubber stopper into the flask. After the penny has stopped reacting, pour the solution into a large beaker of water and observe the color.

   Note: The Instructor should release the gas, NO_2, under a fume hood. Students should try to explain the brown smog and the blue solution.

4. *Here and Gone.* Measure about 100 mL of 0.1 *M* copper(II) sulfate into a tall glass cylinder. Add about 1 mL of 6 *M* ammonium hydroxide solution to the cylinder and observe the reaction. Add an additional 20 mL of 6 *M* ammonium hydroxide to the cylinder. Observe the reaction and propose a hypothesis for the observations.

   Note: Students should try to explain the formation of the blue-white solid and its disappearance to form a deep violet solution.

5. *Water Hazard.* Grind about 3 g of ammonium nitrate in a mortar with a pestle. Empty the powder into an evaporating dish; sprinkle fresh zinc dust over the mixture. Carefully spray distilled water from a wash bottle onto the chemicals.

   Note: The reaction is exothermic and should be performed with CAUTION under a fume hood. Students should try to explain the intense reaction.

B. Student Experiments

Students perform each of the following as a chemical demonstration. Students will record observations in the Data Table and propose a hypothesis to explain their observation.

1. *Disappearing Blue*. Observe the clear solution in the 1000-mL Florence flask. Shake the flask once with your thumb firmly holding the stopper. Wait several seconds; repeat the procedure and record your observations.

   Note: Do not discard the blue solution in the Florence flask, as it can be used repeatedly as a chemical demonstration.

RECYCLE Chemical Waste

2. *Hot and Cold*. Using a spatula, add a small amount of ammonium chloride into one test tube and a small amount of calcium chloride into a second. Half-fill the test tubes with distilled water. Place your hand around the bottom of each test tube and record your observations.

3. *Active and Unreactive*. Half-fill a 250-mL beaker with distilled water. Place an iron nail and a small piece of calcium metal in the water and record your observations.

4. *Copper Nails*. Half-fill a 100-mL beaker with copper sulfate solution. Place an iron nail in the solution. Wait several minutes, remove the nail, and record your observations.

5. *Mirror Images*. Given a ball-and-stick model kit, construct the model shown in Figure 1.2. The letter abbreviations on the balls are as follows: B—black, Y—yellow, O—orange, R—red, and G—green.

   Construct a model identical to the first model. On the second model, switch the positions of the red and yellow balls. Can the two models now be superimposed? Are the two models identical? Diagram each model in the Data Table.

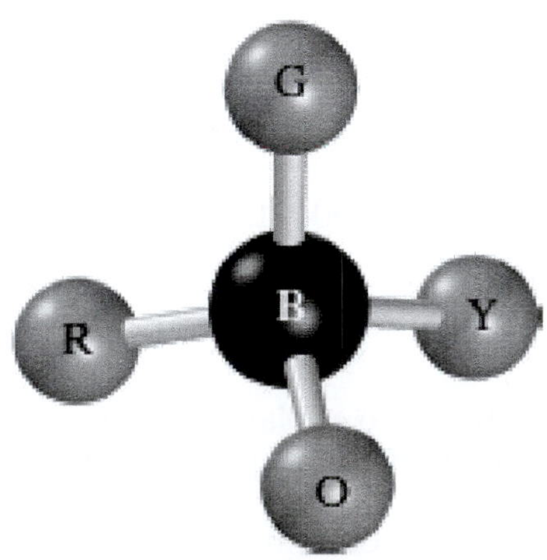

Figure 1.2 Ball-and-Stick Model The illustration shows a molecular model that has a nonidentical mirror image.

EXPERIMENT 1 NAME ____________________

DATE ______________ SECTION ____________________

*PRELABORATORY ASSIGNMENT**

1. Provide the key term that corresponds to each of the following definitions.

__________ (a) the branch of science that studies the composition and properties of matter

__________ (b) the methodical exploration of nature and logical explanation of the observations

__________ (c) a systematic investigation that involves performing an experiment, proposing a hypothesis, testing the hypothesis, and stating a theory or law

__________ (d) a scientific procedure for collecting data and recording observations under controlled conditions

__________ (e) a tentative proposal of a scientific principle that attempts to explain the meaning of the data collected in an experiment

__________ (f) an extensively tested proposal of a scientific principle that explains the behavior of nature

Key Terms: chemistry, experiment, hypothesis, science, scientific method, theory

2. What is the name of the following lab equipment?

(a)

(b)

(c)

(d)

(e)

(f)

* *Answers in Appendix J*

3. State whether each of the following laboratory instructions is *true* or *false*.

(a) ________ Do the Prelaboratory Assignment *after* coming to laboratory and check your answers in **Appendix J**.

(b) ________ Record your observations directly in the Data Table; do not record data on loose scraps of paper.

(c) ________ Dispose of chemicals in a designated container for chemical waste.

(d) ________ Dispose of broken glass in a designated container for broken glass.

(e) ________ Use distilled water when performing an experiment.

(f) ________ Clean glassware with tap water and rinse with distilled water.

(g) ________ Never place chemicals directly on a balance pan.

(h) ________ Never place hot objects on a balance pan.

(i) ________ Allow heated glassware to cool before touching.

(j) ________ Clean your lab station and equipment after completing the experiment.

4. What should you do if any chemical comes in contact with your skin?

5. What safety precautions must be observed in this experiment?

EXPERIMENT 1 NAME ______________________

DATE ______________ SECTION ______________________

DATA TABLE

A. Instructor Demonstrations

1. *Cold Heat*

Observation	Hypothesis

2. *Black Foam*

Observation	Hypothesis

RECYCLE Chemical Waste

3. *Copper Smog*

Observation	Hypothesis

4. *Here and Gone*

Observation	Hypothesis

5. *Water Hazard*

Observation	Hypothesis

B. Student Experiments

1. *Disappearing Blue*

Observation	Hypothesis

2. *Hot and Cold*

Observation	Hypothesis

RECYCLE
Chemical
Waste

3. *Active and Unreactive*

Observation	Hypothesis

4. *Copper Nails*

Observation	Hypothesis

5. *Mirror Images*

Observation	Hypothesis

EXPERIMENT 1 NAME ______________________

DATE ______________ SECTION ______________________

POSTLABORATORY ASSIGNMENT

1. State whether each of the following laboratory safety precautions is *true* or *false*.

   (a) __________ Wear safety goggles in the laboratory.

   (b) __________ Wear closed-toe shoes in the laboratory.

   (c) __________ Do not bring food or drink into the laboratory.

   (d) __________ Note the location of the fire extinguisher(s) in the laboratory.

   (e) __________ Note the location of the first-aid equipment in the laboratory.

   (f) __________ Do not perform unauthorized experiments.

   (g) __________ Waft a gas toward your nose when detecting an odor.

   (h) __________ Perform experiments that produce a gas under a fume hood.

   (i) __________ When heating a test tube, point the open end in a safe direction.

   (j) __________ Always pour an acid into water—not water into acid.

   (k) __________ Clean up broken glass immediately.

   (l) __________ Do not use an organic liquid near an open flame in the laboratory.

   (m) __________ If you contact a chemical, wash with water and notify the Instructor.

   (n) __________ Notify the Instructor immediately in the event of an accident.

2. Which of the following chemicals should be handled carefully in the laboratory?

   (a) acids

   (b) bases

   (c) alcohol

   (d) distilled water

3. (optional) You are given nine pennies. One penny was minted in 1980 and the other eight pennies were minted after 1982. The 1980 penny weighs 3.0 grams; the other pennies have less copper and weigh only 2.5 grams. Assuming the mint dates are illegible, devise a method using the balance shown to determine the heavier 1980 penny in only two trials.

EXPERIMENT

2

Instrumental Measurements

OBJECTIVES

- To obtain measurements of length, mass, volume, and temperature.
- To determine the mass and volume of an unknown rectangular solid.
- To gain proficiency in using the following instruments: metric rulers, balances, graduated cylinder, and thermometer.

DISCUSSION

The **metric system** uses a basic set of units and prefixes. The basic unit of length is the meter, the basic unit of mass is the gram, and the basic unit of volume is the liter. Metric prefixes make these basic units larger or smaller by powers of 10. For example, a kilometer is a thousand times longer than a meter, and a meter is a thousand times longer than a millimeter. In the laboratory, the most common unit of length is **centimeter** (symbol **cm**), the most common unit of mass is **gram** (symbol **g**), and the most common unit of volume is **milliliter** (symbol **mL**).

Scientific instruments have evolved to a high state of sensitivity. However, it is not possible to make an exact measurement. The reason is that all instruments possess a degree of **uncertainty**—no matter how sensitive. The uncertainty is indicated by the significant digits in the measurement. For example, a metric ruler may measure length to the nearest tenth of a centimeter (± 0.1 cm). A different metric ruler may measure length to the nearest five hundredths of a centimeter (± 0.05 cm). The measurement with the least uncertainty (± 0.05 cm) is more precise.

In this experiment, we will use several instruments. We will make measurements of mass with balances having progressively greater sensitivity. A decigram balance is so named because the uncertainty is one-tenth of a gram (± 0.1 g). The uncertainty of a centigram balance is one-hundredth of a gram (± 0.01 g), and the uncertainty of a milligram balance is one-thousandth of a gram (± 0.001 g).

We will make length measurements using two metric rulers that differ in their uncertainty. METRIC RULER A is calibrated in 1-cm divisions and has an uncertainty of ± 0.1 cm. METRIC RULER B has 0.1-cm subdivisions and an uncertainty of ± 0.05 cm. Thus, METRIC RULER B has less uncertainty than METRIC RULER A. The following examples demonstrate measurement of length utilizing the two different metric rulers.

Example Exercise 2.1 • Measuring Length with Metric Ruler A

A copper rod is measured with the metric ruler shown below. What is the length of the rod?

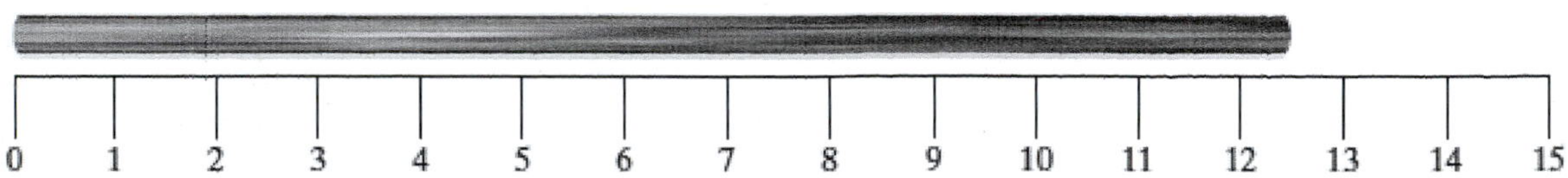

METRIC RULER A *(Estimate to a tenth of a division ±0.1 cm)*

Solution: Each division represents one centimeter. The end of the rod lies between the 12th and 13th divisions. We can estimate to a tenth of a division (± 0.1 cm). Since the end of the rod lies about five-tenths past 12, we can estimate the length as

$$12 \text{ cm} + 0.5 \text{ cm} = 12.5 \text{ cm}$$

Example Exercise 2.2 • Measuring Length with Metric Ruler B

The same copper rod is measured with the metric ruler shown below. What is the length of the rod?

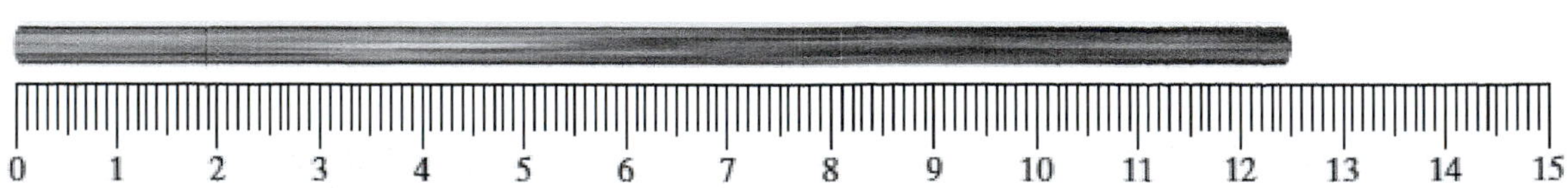

METRIC RULER B *(Estimate to a half of a subdivision ±0.05 cm)*

Solution: Note that this ruler is divided into centimeters that are subdivided into tenths of centimeters. The end of the rod lies between the 12th and 13th divisions and between the 5th and 6th subdivisions. Thus, the length is between 12.5 cm and 12.6 cm.

We can estimate the measurement more precisely. A subdivision is too small to divide into ten parts, but we can estimate to half of a subdivision (± 0.05 cm). The length is 12 cm + 0.5 cm + 0.05 cm = 12.55 cm.

To test your skill in making metric measurements, you will determine the mass and volume of an unknown rectangular solid. The volume of a rectangular solid is calculated from its length, width, and thickness. The following examples will illustrate.

Example Exercise 2.3 • Calculating Volume of a Rectangular Solid

An unknown rectangular solid was measured with METRIC RULER A, which provided the following: 5.0 cm by 2.5 cm by 1.1 cm. What is the volume of the solid?

Solution: The volume of a rectangular solid equals length times width times thickness.

$$5.0 \text{ cm} \times 2.5 \text{ cm} \times 1.1 \text{ cm} = 13.75 \text{ cm}^3 = 14 \text{ cm}^3$$

In this example, each measurement has two significant digits; thus, the volume has two significant digits. Note the unit of volume is cubic centimeter, cm^3.

Example Exercise 2.4 • Calculating Volume of a Rectangular Solid

The unknown rectangular solid was also measured with METRIC RULER B, which gave the following: 5.00 cm by 2.45 cm by 1.15 cm. What is the volume of the solid?

Solution: The volume of a rectangular solid equals length times width times thickness.

$$5.00 \text{ cm} \times 2.45 \text{ cm} \times 1.15 \text{ cm} = 14.0875 \text{ cm}^3 = 14.1 \text{ cm}^3$$

In this example, each measurement has three significant digits; thus, the volume has three significant digits.

We can measure the volume of a liquid using a graduated cylinder. If we carefully examine the 100-mL graduated cylinder shown in Figure 2.1, we notice that it is marked in 10-mL intervals, and each interval has ten subdivisions. Therefore, each subdivision equals one milliliter. If we estimate to half of a subdivision, the uncertainty is ± 0.5 mL.

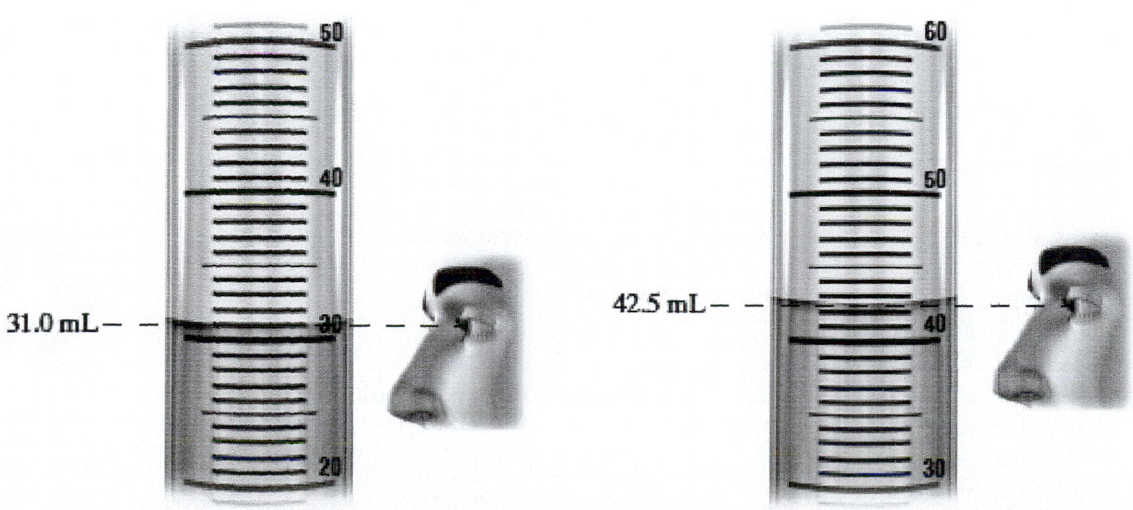

Figure 2.1 Graduated Cylinder Example readings using proper eye position and recording the bottom of the **meniscus** to half a subdivision (± 0.5 mL).

We can measure temperature using a Celsius thermometer. If we examine the thermometer shown in Figure 2.2, we notice that it is marked in 10 °C intervals that have ten subdivisions. Thus, each subdivision equals one degree Celsius. If we estimate to half of a subdivision, the temperature measurement has an uncertainty of ± 0.5 °C.

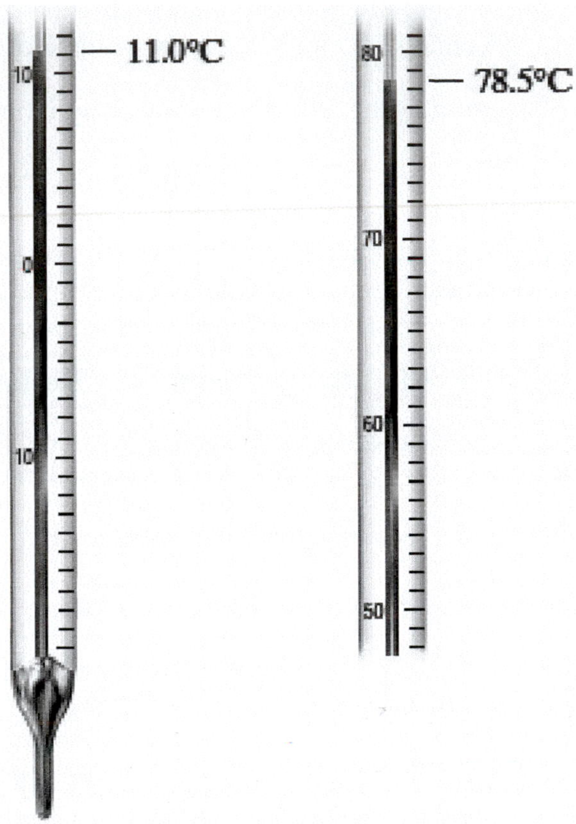

Figure 2.2 Celsius Thermometer Example readings using a Celsius thermometer and recording the top of the liquid to half a subdivision (± 0.5 °C).

EQUIPMENT and CHEMICALS

- 13 x 100 mm test tubes (3)
- watchglass
- evaporating dish
- crucible & cover
- 125-mL Erlenmeyer flask
- 100-mL graduated cylinder
- dropper pipet
- 250-mL beaker with ice
- 150-mL beaker
- 110 °C thermometer
- ring stand & ring
- ring stand & ring
- wire gauze
- decigram balance
- centigram balance
- milligram balance
- unknown rectangular solid

PROCEDURE

A. Length Measurements

1. Measure the length of a 13 x 100 mm test tube with each of the following: (a) METRIC RULER A, and (b) METRIC RULER B.

   Note: Refer to METRIC RULER A instructions in **Example Exercise 2.1**.
   Refer to METRIC RULER B instructions in **Example Exercise 2.2**.

2. Measure the diameter of a watchglass with each of the following: (a) METRIC RULER A, and (b) METRIC RULER B.

3. Measure the diameter of an evaporating dish (not the spout) with each of the following: (a) METRIC RULER A, and (b) METRIC RULER B.

B. Mass Measurements

1. Determine the mass of an evaporating dish on the following balances: (a) decigram balance, (b) centigram balance, and (c) milligram balance.

2. Determine the mass of a crucible and cover on the following balances: (a) decigram balance, (b) centigram balance, and (c) milligram balance.

3. Determine the mass of a 125-mL Erlenmeyer flask on the following balances: (a) decigram balance, (b) centigram balance, and (c) milligram balance.

   Note: Refer to balance instructions in **Appendices B**, **C**, and **D**.

C. Mass and Volume of an Unknown Solid

1. Obtain a rectangular solid and record the unknown number in the Data Table. Find the mass of the unknown rectangular solid using each of the following: (a) a decigram balance, (b) a centigram balance, and (c) a milligram balance.

2. Measure the length, width, and thickness of the rectangular solid unknown using METRIC RULER A shown in Example Exercise 2.1. Calculate the volume.

3. Measure the length, width, and thickness of the rectangular solid unknown using METRIC RULER B shown in Example Exercise 2.2. Calculate the volume.

D. Volume Measurements

1. Fill a 100-mL graduated cylinder with water. Adjust the bottom of the meniscus to the full mark with a dropper pipet. Record the volume as 100.0 mL.

2. Fill a 13 x 100 mm test tube with water from the graduated cylinder. Record the new volume in the graduated cylinder (± 0.5 mL).

   Note: Refer to the graduated cylinder instructions in **Figure 2.1**.

3. Fill a second test tube with water. Record the volume in the graduated cylinder.

E. Temperature Measurements

1. Record the temperature in the laboratory using a Celsius thermometer (± 0.5 °C).

2. Half-fill a 250-mL beaker with ice and water. Hold the thermometer in the ice water and record the coldest observed temperature (± 0.5 °C).

3. Half-fill a 150-mL beaker with distilled water. Support the beaker on a ring stand with a wire gauze as shown in Figure 2.3. Heat the water to boiling and shut off the burner. Place the thermometer in the boiling water and record the temperature (± 0.5 °C).

   Note: Refer to the laboratory burner instructions in **Appendix A**.

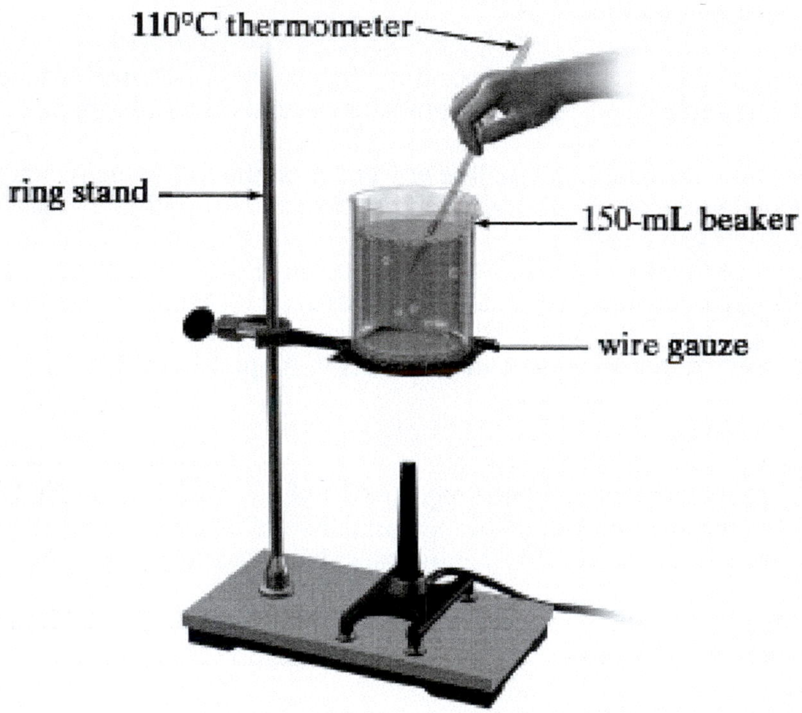

Figure 2.3 Apparatus for Boiling Water To obtain an accurate temperature measurement, do not allow the thermometer to touch the hot glass beaker.

EXPERIMENT 2 NAME ______________________

DATE ______________ SECTION ______________________

*PRELABORATORY ASSIGNMENT**

1. Provide the key term that corresponds to each of the following definitions.

_________ (a) a decimal system of measurement using prefixes and a basic unit to express length, mass, and volume

_________ (b) a metric unit of length

_________ (c) a metric unit of mass

_________ (d) a metric unit of volume

_________ (e) the clear lens at the surface of a liquid inside a graduated cylinder

_________ (f) the degree of inexactness in an instrumental measurement

Key Terms: centimeter (cm), gram (g), meniscus, metric system, milliliter (mL), uncertainty

2. State the length measurement indicated on each of the following metric rulers.

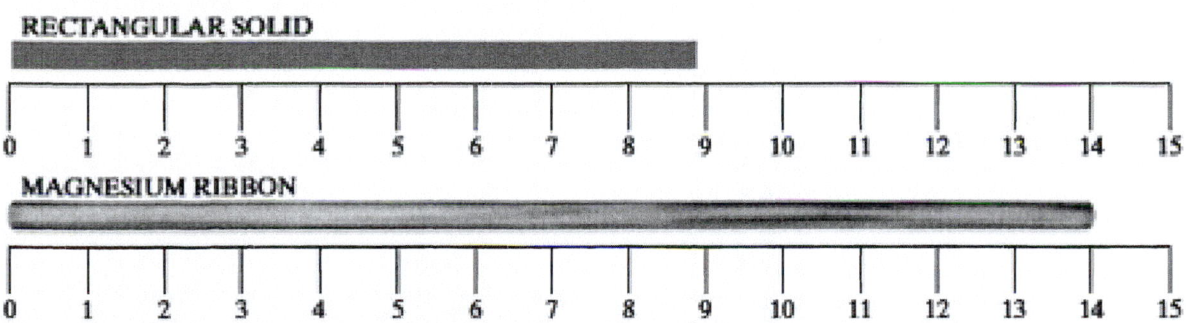

3. A rectangular solid measures 5.0 cm by 2.5 cm by 1.5 cm. Refer to Example Exercise 2.3 and show the calculation for volume of the rectangular solid.

4. State the length measurement indicated on each of the following metric rulers.

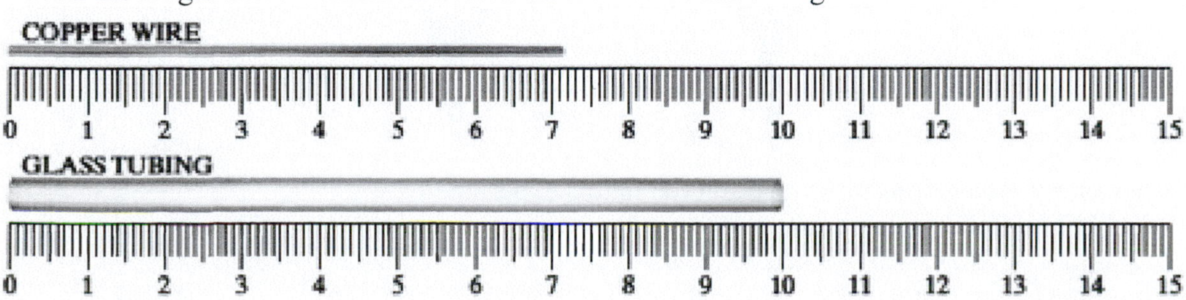

* *Answers in Appendix J*

5. A rectangular solid measures 5.05 cm by 2.45 cm by 1.50 cm. Refer to Example Exercise 2.4 and show the calculation for volume of the rectangular solid.

6. State the volume measurement indicated by each of the following graduated cylinders.

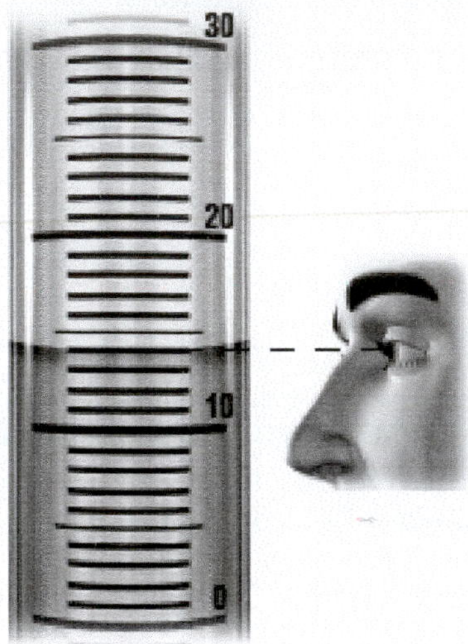

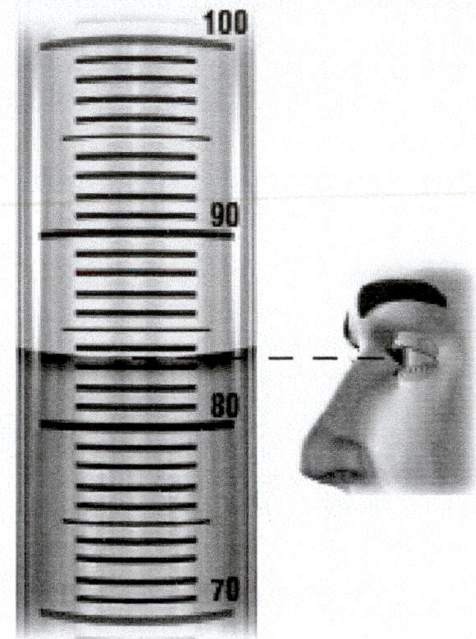

7. State the temperature measurement indicated by each of the following Celsius thermometers.

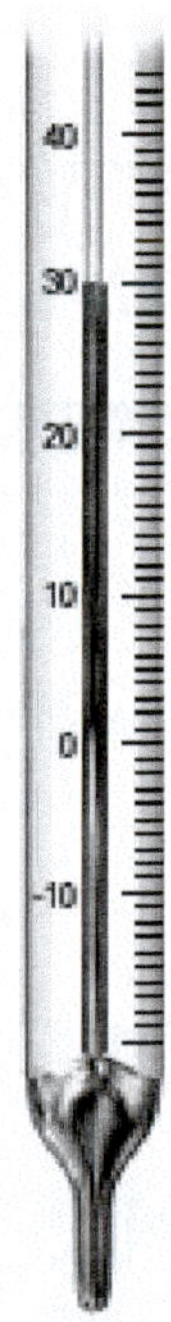

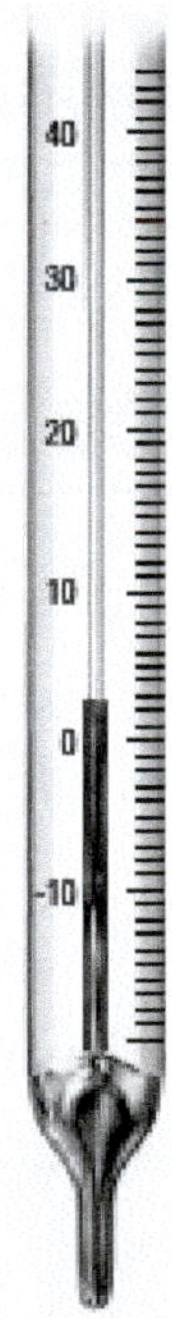

8. What safety precautions must be observed in this experiment?

EXPERIMENT 2 NAME ______________________________

DATE ____________________ SECTION ______________________________

DATA TABLE

A. Length Measurements

length of a 13 x 100 mm test tube

METRIC RULER A	______________	cm
METRIC RULER B	______________	cm

diameter of a watchglass

METRIC RULER A	______________	cm
METRIC RULER B	______________	cm

diameter of an evaporating dish

METRIC RULER A	______________	cm
METRIC RULER B	______________	cm

B. Mass Measurements

mass of an evaporating dish

decigram balance	______________	g
centigram balance	______________	g
milligram balance	______________	g

mass of a crucible and cover

decigram balance	______________	g
centigram balance	______________	g
milligram balance	______________	g

mass of a 125-mL Erlenmeyer flask

decigram balance	______________	g
centigram balance	______________	g
milligram balance	______________	g

C. Mass and Volume of an Unknown Solid **UNKNOWN #** ________

mass of unknown solid

decigram balance ____________ g

centigram balance ____________ g

milligram balance ____________ g

volume of unknown solid (METRIC RULER A)

length of solid ____________ cm

width of solid ____________ cm

thickness of solid ____________ cm

Show the calculation for the volume of the rectangular solid (see Example Exercise 2.3).

____________ cm^3

volume of unknown solid (METRIC RULER B)

length of solid ____________ cm

width of solid ____________ cm

thickness of solid ____________ cm

Show the calculation for the volume of the rectangular solid (see Example Exercise 2.4).

____________ cm^3

D. Volume Measurements

volume of water in a graduated cylinder ____________ mL

volume minus one test tube of water ____________ mL

volume minus two test tubes of water ____________ mL

E. Temperature Measurements

room temperature ____________ °C

melting point temperature of ice ____________ °C

boiling point temperature of water ____________ °C

EXPERIMENT 2 NAME ____________________

DATE ______________ SECTION ____________________

POSTLABORATORY ASSIGNMENT

1. State the basic unit in the metric system for each of the following.

(a) length ________ (b) mass ________

(c) volume ________ (d) temperature ________

2. State a common laboratory instrument for measuring each of the following.

(a) diameter of a beaker ________ (b) mass of a sample ________

(c) volume of water ________ (d) temperature of air ________

3. State the metric unit associated with each of the following instruments.

(a) metric ruler ________ (b) balance ________

(c) graduated cylinder ________ (d) thermometer ________

4. Select the measurement that is consistent with the uncertainty of each instrument.

(a) METRIC RULER A: 5 cm, 5.0 cm, 5.00 cm ________

(b) METRIC RULER B: 5 cm, 5.0 cm, 5.00 cm ________

(c) decigram balance: 5.0 g, 5.00 g, 5.000 g ________

(d) centigram balance: 5.0 g, 5.00 g, 5.000 g ________

(e) milligram balance: 5.0 g, 5.00 g, 5.000 g ________

(f) graduated cylinder: 5 mL, 5.0 mL, 5.00 mL ________

(g) Celsius thermometer: 5 °C, 5.0 °C, 5.00 °C ________

5. State the uncertainty (for example, ± 0.5 cm) in each of the following measurements.

(a) 25.00 cm ________ (b) 25.000 g ________

(c) 25.0 mL ________ (d) 25.0 °C ________

6. State the number of significant digits in each of the following measurements.

(a) 5.00 cm ________ (b) 0.50 cm ________

(c) 0.500 g ________ (d) 0.005 g ________

(e) 50.0 mL ________ (f) 5.5 mL ________

(g) 50.5 °C ________ (h) –0.5 °C ________

7. Perform the indicated math operation and round off the answer to the proper significant digits.

(a)
$$\begin{array}{r} 50.511\ \text{g} \\ +\ \underline{10.25\ \ \text{g}} \end{array}$$

(b)
$$\begin{array}{r} 97.5\ \ \ \text{g} \\ -\ \underline{95.826\ \text{g}} \end{array}$$

8. Perform the indicated math operation and round off the answer to the proper significant digits.

(a) (5.15 cm) (2.25 cm) (1.0 cm)

(b) $\dfrac{15.15\ \text{cm}^3}{12.0\ \text{cm}^2}$

9. Explain why you round off the numbers in a calculator display after addition, subtraction, multiplication, or division of measurements.

10. (optional) A platinum cylinder has a mass of 1.000 kg, a diameter of 3.90 cm, and a height of 3.90 cm. What is the volume of the cylinder in cubic centimeters? The volume of a cylinder equals $\pi d^2 h/4$, where π is 3.14, d is the diameter, and h is the height.

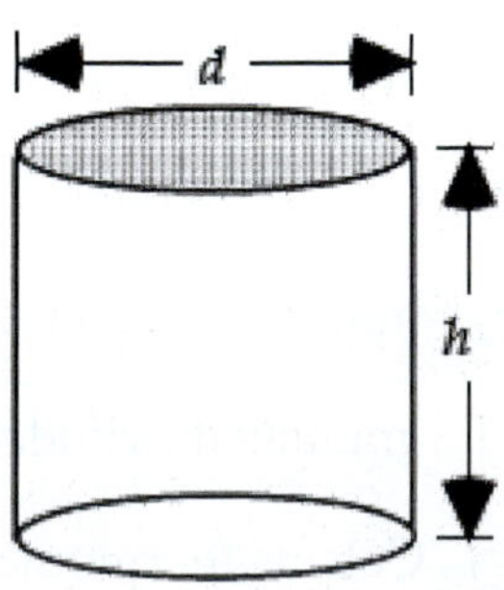

EXPERIMENT

3

Density of Liquids and Solids

OBJECTIVES

- To observe the relative density of common liquids and solids.
- To determine the density of water, an unknown liquid, a rubber stopper, and an unknown rectangular solid.
- To determine the thickness of a piece of aluminum foil using the density concept.
- To gain proficiency in performing the following experimental procedures: pipetting a liquid, weighing by difference, and determining a volume by displacement.

DISCUSSION

Density is a physical property of liquids and solids. We can define **density** (symbol ***d***) as the amount of mass in a given volume. To determine the density of a solid experimentally, we must measure the mass of the solid using a balance. To determine the mass of a liquid, we use an indirect technique called **weighing by difference** (Figure 3.1). First, we weigh a flask empty. Second, we add a given volume of liquid into the flask and reweigh. The mass of the liquid is found by subtracting the first mass reading from the second mass reading.

After collecting the experimental data, we can calculate density by dividing the mass by the volume. It is important, however, that we attach the proper units to the calculated value. The density of liquids and solids is usually expressed in grams per milliliter (g/mL) or grams per cubic centimeter (g/cm^3). Since 1 mL = 1 cm^3, the numerical value for density in g/mL and g/cm^3 is identical. For example, the density of water may be expressed as 1.00 g/mL or 1.00 g/cm^3.

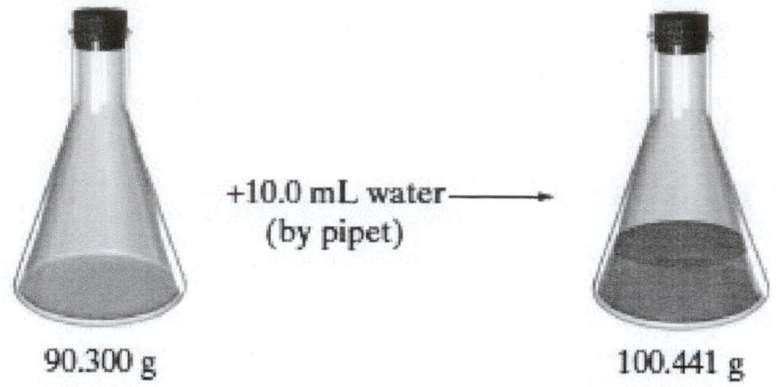

Figure 3.1 Weighing by Difference The mass of the liquid is found by the difference in masses: 100.441 g – 90.300 g = 10.141 g.

Example Exercise 3.1 • Density of a Liquid

A 10.0-mL sample of water is pipetted into a flask. The mass of water, 10.141 g, is found after weighing by difference (see Figure 3.1). Calculate the density of water.

Solution: Dividing the mass of water by volume, we have

$$\frac{10.141 \text{ g}}{10.0 \text{ mL}} = 1.01 \text{ g/mL}$$

We round the answer to three significant digits because there are only three digits in the denominator. In this example, the calculated value, 1.01 g/mL, agrees closely with the theoretical value, 1.00 g/mL. The slight discrepancy is due to experimental error.

The volume of an irregular object can be found indirectly from the amount of water it displaces. This technique is called **volume by displacement**. For example, the volume of a rubber stopper can be determined as shown in Figure 3.2. The initial reading of water in the graduated cylinder is observed. The stopper is introduced into the graduated cylinder and the final reading is recorded. The difference between the initial and final readings corresponds to the volume of water displaced. The volume of water displaced is equal to the volume of the rubber stopper.

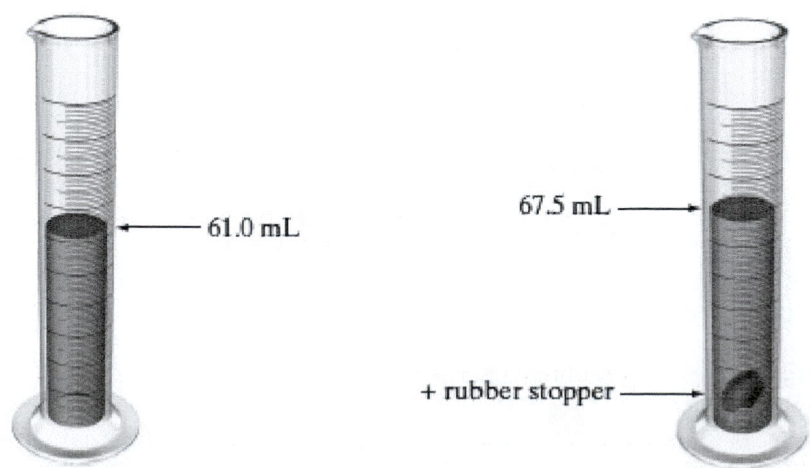

Figure 3.2 Volume by Displacement The volume of the rubber stopper is found by the increase in volume: 67.5 mL – 61.0 mL = 6.5 mL.

Example Exercise 3.2 • Density of a Rubber Stopper

A rubber stopper weighing 8.453 g displaces 6.5 mL of water in a graduated cylinder (Figure 3.2). What is the density of the rubber stopper?

Solution: Dividing the mass of the rubber stopper by its volume, we have

$$\frac{8.453\ \text{g}}{6.5\ \text{mL}} = 1.3\ \text{g/mL}$$

In this example, the volume has two significant digits. Thus, the density of the rubber stopper is limited to two digits.

We will also determine the density of a solid. The volume of any solid object with regular dimensions can be found by calculation. For example, the volume of a rectangular solid object is calculated by multiplying its length times its width times its thickness.

Example Exercise 3.3 • Density of a Rectangular Solid

The mass of an unknown rectangular block is 139.443 g. If the block measures 5.00 cm by 2.55 cm by 1.25 cm, what is its density?

Solution: First, we calculate the volume of the rectangular block.

$$5.00\ \text{cm} \times 2.55\ \text{cm} \times 1.25\ \text{cm} = 15.9\ \text{cm}^3$$

Second, we find the density of the unknown rectangular solid.

$$\frac{139.443\ \text{g}}{15.9\ \text{cm}^3} = 8.77\ \text{g/cm}^3$$

The thickness of aluminum foil is too thin to measure with a ruler. However, we can find the thickness of the foil indirectly. Given the mass and density of the foil, we can calculate the volume. From the volume, length, and width of the foil, we can calculate the thickness.

Example Exercise 3.4 • Thickness of an Aluminum Foil

A piece of aluminum foil has a mass of 0.450 g and measures 10.75 cm by 10.10 cm. Given the density of aluminum, 2.70 g/cm^3, calculate the thickness of the foil.

Solution: To calculate the thickness of the foil, we must first find the volume. The volume can be calculated using density as a unit factor.

$$0.450\ \cancel{\text{g}} \times \frac{1\ \text{cm}^3}{2.70\ \cancel{\text{g}}} = 0.167\ \text{cm}^3$$

The thickness is found after dividing the volume by its length and width.

$$\frac{0.167\ \text{cm}^3}{(10.75\ \text{cm})(10.10\ \text{cm})} = 0.00154\ \text{cm} \quad (1.54 \times 10^{-3}\ \text{cm})$$

EQUIPMENT and CHEMICALS

A. Instructor Demonstration

- tall glass cylinder
- corn syrup
- mineral oil
- glass marble
- rubber stopper
- ice
- cork

B–F. Student Experiments

- 125-mL Erlenmeyer flask with rubber stopper to fit
- 150-mL beaker
- 100-mL beaker
- 10-mL pipet & bulb
- 100-mL graduated cylinder
- #2 rubber stopper
- unknown liquids
- unknown rectangular solids
- aluminum foil, ~ 10 x 10 cm rectangle

PROCEDURE

A. Instructor Demonstration – Density

RECYCLE Chemical Waste

1. Add ~100 mL of corn syrup into a tall glass cylinder.
2. Slowly add ~200 mL of water into the cylinder.
3. Slowly add ~100 mL of mineral oil into the cylinder.
4. Slowly slide a glass marble into the tall glass cylinder.
5. Slowly slide a rubber stopper into the cylinder.
6. Slowly slide a piece of ice into the cylinder.
7. Drop a cork into the cylinder.

B. Density of Water

The Instructor may demonstrate how to *condition* a pipet and *transfer* a sample liquid.

1. Weigh a 125-mL Erlenmeyer flask fitted with a rubber stopper.
2. Half-fill a 150-mL beaker with distilled water, and then pipet a 10.0-mL sample into the 125-mL flask (see **Appendix E**).
3. Reweigh the flask and stopper, and determine the mass of water by difference.
4. Repeat a second trial for the density of the water.

   Note: It is not necessary to dry the flask between trials because the 10.0-mL sample of water is weighed by difference.

5. Calculate the density of water for each trial, and report the average value for both trials.

C. Density of an Unknown Liquid

1. Obtain about 25 mL of an unknown liquid in a 100-mL beaker. Record the unknown number in the Data Table.

2. Weigh a 125-mL Erlenmeyer flask fitted with a rubber stopper.

3. Condition a pipet with unknown liquid, and transfer a 10.0-mL sample into the flask.

RECYCLE
Chemical
Waste

4. Reweigh the flask and stopper, and determine the mass of liquid by difference.

5. Repeat a second trial for the density of the unknown liquid.

6. Calculate the density of the unknown liquid and report the average value for both trials.

D. Density of a Rubber Stopper

1. Weigh a dry #2 rubber stopper.

2. Half-fill a 100-mL graduated cylinder with water. Observe the bottom of the meniscus and estimate the volume to ±0.5 mL (see Figure 3.3).

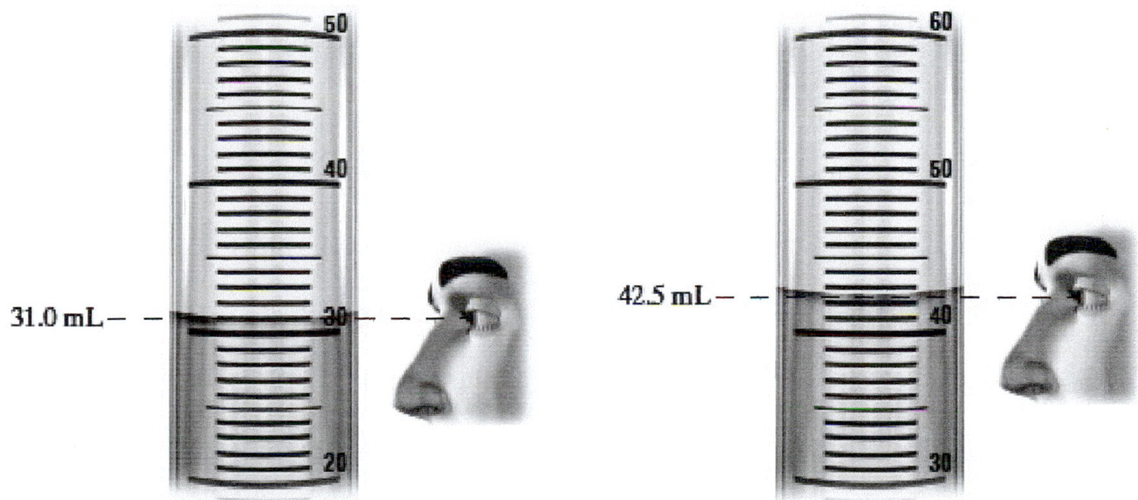

Figure 3.3 Volume by Displacement Use proper eye position and record the bottom of the **meniscus** to half a subdivision (± 0.5 mL).

3. Tilt the graduated cylinder, and let the stopper slowly slide into the water. Observe the new water level, and calculate the volume by displacement for the stopper.

4. Repeat a second trial for the density of the rubber stopper.

5. Calculate the density of the rubber stopper and report the average value for both trials.

E. Density of an Unknown Solid

1. Obtain a rectangular solid, and record the unknown number in the Data Table.
2. Weigh the unknown solid and record the mass.
3. Measure and record the length, width, and thickness of the unknown rectangular solid, using the metric ruler in Figure 3.4.

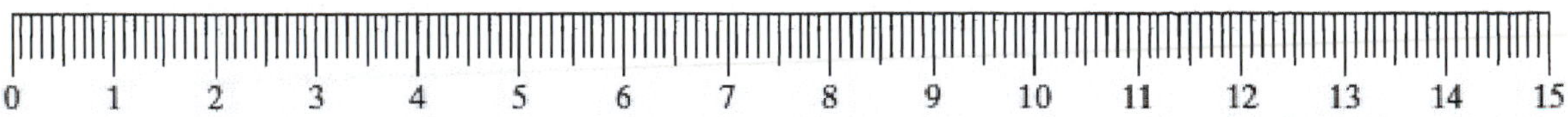

(One subdivision = 0.1 cm)

Figure 3.4 Metric Ruler The uncertainty of the measurement is ±0.05 cm.

4. Calculate the volume of the unknown rectangular solid.
5. Repeat a second trial for the volume of the unknown solid using a different balance and the metric ruler in Figure 3.4.

F. Thickness of Aluminum Foil

1. Obtain a rectangular piece of aluminum foil.
2. Measure the length and width of the foil (refer to the metric ruler Figure 3.4).
3. Weigh the aluminum foil and record the mass in the Data Table.
4. Calculate the volume and thickness of the aluminum foil ($d = 2.70$ g/cm^3).

EXPERIMENT 3 NAME ______________________

DATE ______________ SECTION ______________________

*PRELABORATORY ASSIGNMENT**

1. Provide the key term that corresponds to each of the following definitions.

_________ (a) the amount of mass in a unit volume of matter; for example, 1.00 g/mL

_________ (b) a procedure for obtaining the mass of a sample by first weighing a container and then weighing the container with the sample

_________ (c) the degree of inexactness in an instrumental measurement

_________ (d) to rinse glassware (e.g., a pipet) with a sample liquid to avoid dilution by water on the inside surface

_________ (e) the clear lens at the surface of a liquid inside a graduated cylinder

_________ (f) determining volume of a sample by measuring the volume of water displaced

Key Terms: condition, density, meniscus, uncertainty, volume by displacement, weighing by difference

2. A 10.0-mL sample of liquid is pipetted into a 125-mL flask with stopper. The mass of liquid is found to be 7.988 g. Refer to Example Exercise 3.1 and calculate the density of the liquid.

3. State the volume of liquid shown in each of the following graduated cylinders.

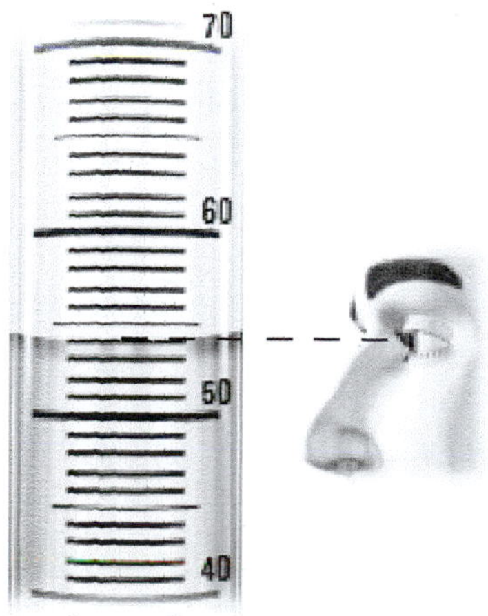

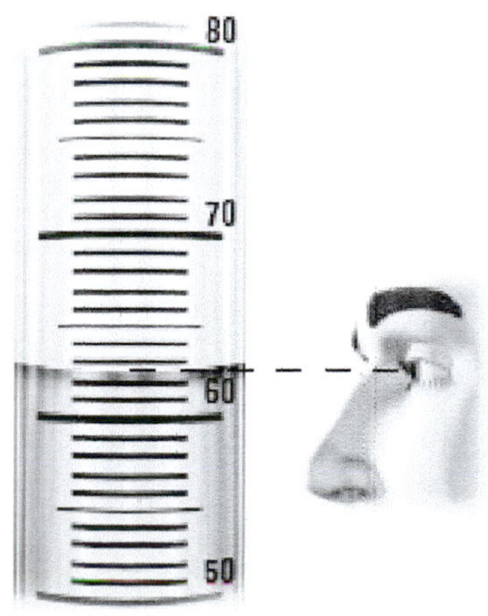

---

* *Answers in Appendix J*

4. A rubber stopper has a mass of 7.452 g and displaces 6.0 mL of water in a graduated cylinder. Refer to Example Exercise 3.2 and calculate the density of the rubber stopper.

5. State the length shown for each of the following rectangular solids.

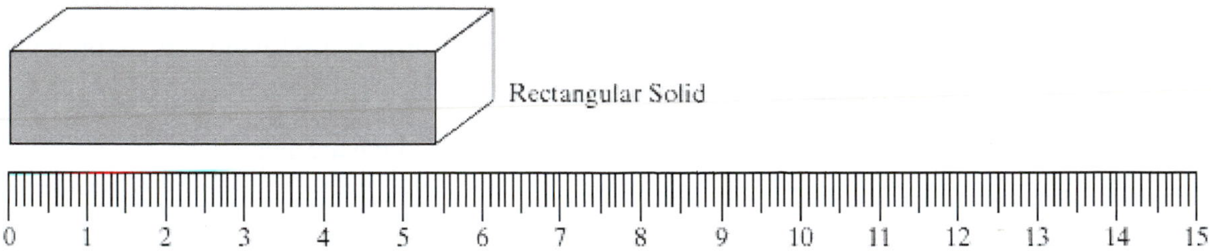

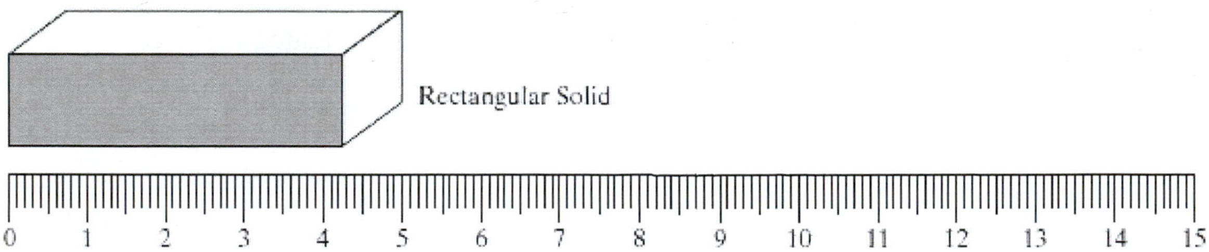

6. An unknown rectangular solid has a mass of 140.417 g and measures 5.05 cm by 2.50 cm by 1.25 cm. Refer to Example Exercise 3.3 and calculate the density of the unknown solid.

7. An aluminum foil weighs 0.465 g and measures 10.10 cm by 10.05 cm. Given the density of aluminum, 2.70 g/cm^3, refer to Example Exercise 3.4 and calculate the thickness of the foil.

8. What safety precautions must be observed in this experiment?

EXPERIMENT 3 NAME ______________________

DATE ______________ SECTION ______________________

DATA TABLE

A. Instructor Demonstration – Density

- *corn syrup*
- *water*
- *mineral oil*
- *glass marble*
- *rubber stopper*
- *ice*
- *cork*

Identify the liquids (**L_1**, **L_2**) and solids (**S_1**, **S_2**, **S_3**, **S_4**) in the tall glass cylinder.

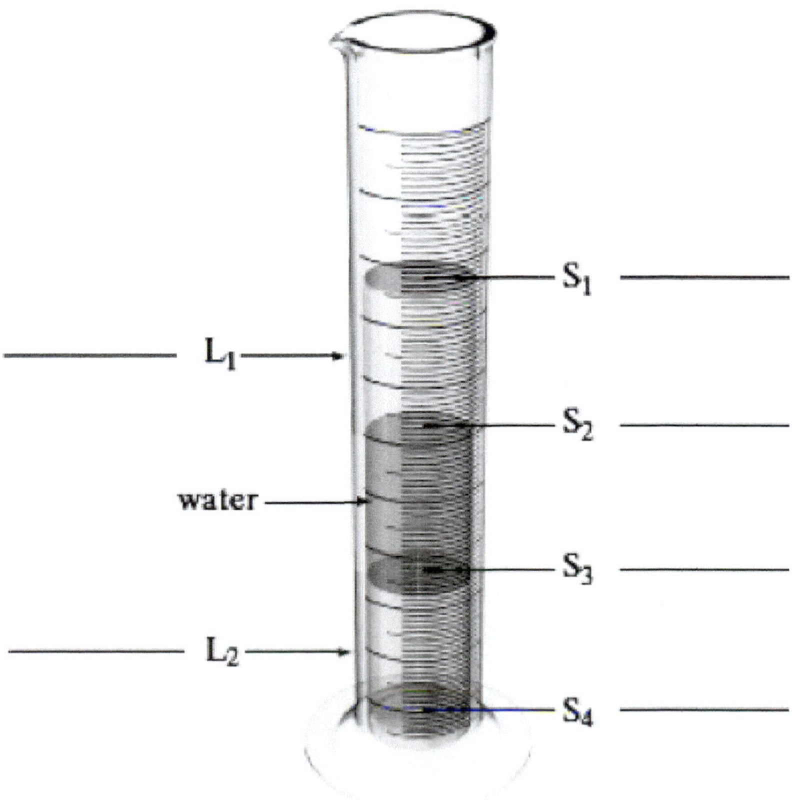

RECYCLE
Chemical
Waste

B. Density of Water

mass of flask and stopper + water	____________ g	____________ g
mass of flask and stopper	____________ g	____________ g
mass of water	____________ g	____________ g
volume of water	____________ mL	____________ mL

Show the calculation for the density of water for trial 1 (see Example Exercise 3.1).

Density of water	____________ g/mL	____________ g/mL

Average density of water ____________ g/mL

C. Density of an Unknown Liquid **UNKNOWN #** _______

RECYCLE
Chemical
Waste

mass of flask and stopper + liquid	____________ g	____________ g
mass of flask and stopper	____________ g	____________ g
mass of unknown liquid	____________ g	____________ g
volume of unknown liquid	____________ mL	____________ mL

Show the calculation for the density of the unknown liquid for trial 1.

Density of unknown liquid	____________ g/mL	____________ g/mL

Average density of unknown liquid ____________ g/mL

D. Density of a Rubber Stopper

mass of a rubber stopper	____________	g	____________	g
final graduated cylinder reading	____________	mL	____________	mL
initial graduated cylinder reading	____________	mL	____________	mL
volume of rubber stopper	____________	mL	____________	mL

Show the calculation of density for the stopper for trial 1 (see Example Exercise 3.2).

Density of rubber stopper	____________	g/mL	____________	g/mL
Average density of rubber stopper			____________	g/mL

E. Density of an Unknown Solid **UNKNOWN #** ________

mass of solid	____________	g	____________	g
length of solid	____________	cm	____________	cm
width of solid	____________	cm	____________	cm
thickness of solid	____________	cm	____________	cm

Show the calculation for the volume of the unknown for trial 1 (see Example Exercise 3.3).

volume of solid	____________	cm^3	____________	cm^3

Show the calculation for the density of the unknown for trial 1 (see Example Exercise 3.3).

Density of rectangular solid	____________	g/cm^3	____________	g/cm^3
Average density of the solid			____________	g/cm^3

F. Thickness of Aluminum Foil

length of foil ____________ cm

width of foil ____________ cm

mass of foil ____________ g

Show the calculation for the volume of the aluminum foil given the density of aluminum, $d = 2.70\ g/cm^3$ (see Example Exercise 3.4).

Volume of foil ____________ cm^3

Show the calculation for the thickness of the foil in centimeters.

Thickness of foil ____________ cm

EXPERIMENT 3 NAME ____________________

DATE ____________________ SECTION ____________________

POSTLABORATORY ASSIGNMENT

1. Ether floats on water, and water floats on mercury, as shown in the following diagram.

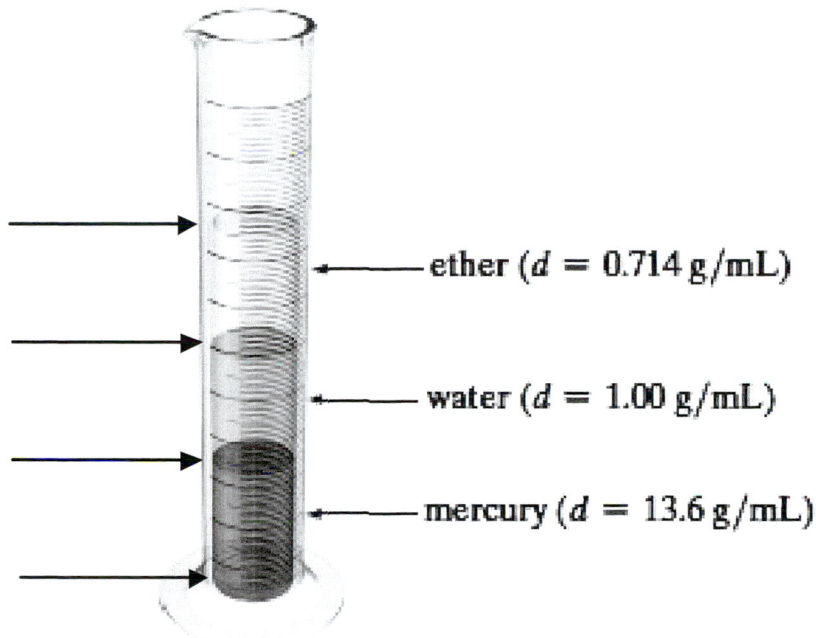

Indicate on the above diagram where each of the following would come to rest after being dropped into the glass cylinder.

(a) a glass marble ($d = 2.95$ g/cm^3)
(b) a platinum ring ($d = 21.45$ g/cm^3)
(c) a lump of coal ($d = 0.83$ g/cm^3)
(d) a champagne cork ($d = 0.19$ g/cm^3)

2. A 250-mL flask and stopper have a mass of 110.525 g. A 50.0-mL sample of gasoline is pipetted into the flask, giving a total mass of 146.770 g. Find the density of the gasoline.

____

3. A piece of green jade has a mass of 26.123 g. If the sample of jade displaces 50.0 mL of water to 57.5 mL in a graduated cylinder, what is the density of the jade?

____

4. A 5.00-cm cube of magnesium has a mass of 217.501 g. What is the density of magnesium metal?

____

5. Aluminum foil is often incorrectly termed tin foil. If the density of tin is 7.28 g/cm^3, what is the thickness of a piece of tin foil that measures 5.70 cm by 4.25 cm and has a mass of 0.655 g?

____

6. (optional) A silver sphere has a mass of 5.492 g and a diameter of 10.0 mm. What is the density of silver metal in grams per cubic centimeter? The volume of a sphere equals $4\pi r^3/3$, where π is 3.14, and r is the radius.

____

EXPERIMENT

4

Freezing Point and Melting Point

OBJECTIVES

- To gain proficiency in constructing a graph and plotting data points.
- To determine the freezing point of a compound from a graph of temperature versus time.
- To determine the melting points of a known and unknown compound.

DISCUSSION

A sample of matter can exist in the solid, liquid, or gaseous state. The **physical state** of a substance depends on the temperature and atmospheric pressure. For example, water can exist as solid ice at temperatures below 0 °C and as gaseous steam above 100 °C.

A **change of state** occurs when there is sufficient heat energy for individual molecules to overcome their attraction for each other. For example, when ice is converted to water, the water molecules in the ice crystal acquire enough energy to become free of each other and move around. Conversely, when water cools to ice, the water molecules lose energy and can no longer move about. Thus, a solid is composed of fixed particles and a liquid has mobile particles.

At the temperature where a liquid changes to a solid, two physical states are present simultaneously. This temperature is referred to as the **freezing point**. Conversely, if a solid changes to a liquid, it is called the **melting point**. Theoretically, the freezing and melting points of a substance occur at the same temperature.

In this experiment, we will melt paradichlorobenzene and then allow the liquid to cool to a solid. We will record the temperature/time relationship, plot the data, and graph a cooling curve. The temperature remains constant as the liquid solidifies. Figure 4.1 shows a typical cooling curve.

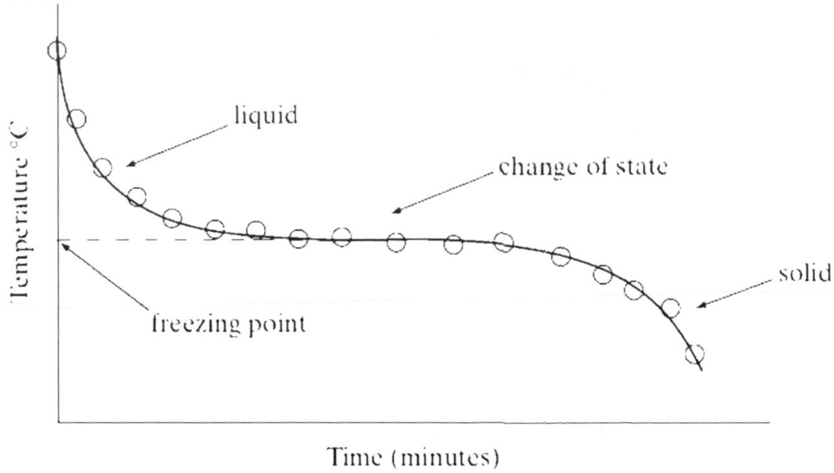

Figure 4.1 Cooling Curve As a liquid cools, it changes state from a liquid to a solid. The freezing point corresponds to the flat plateau portion of the curve.

As the compound cools, crystals begin to form. After a few minutes, the crystals become a solid mass as the liquid changes to a solid. We will plot temperature on the vertical axis, which is called the **ordinate**. We will plot time on the horizontal axis, which is referred to as the **abscissa**. The freezing point of the compound is the temperature corresponding to the flat plateau. The apparatus for determining the cooling curve is shown in Figure 4.2.

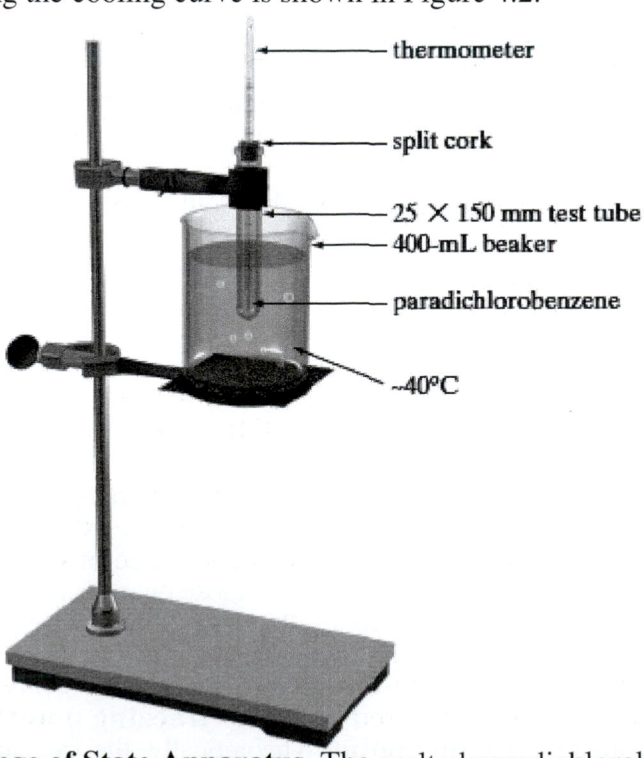

Figure 4.2 Change of State Apparatus The melted paradichlorobenzene is inside a test tube, which in turn, is placed in a beaker of water at ~40 °C.

In the second procedure of this experiment a melting point is determined. A small sample of compound is rapidly heated until it is observed to liquefy. The temperature range over which the compound melts is recorded; for example, 65–75 °C. A second trial is repeated for greater accuracy. The waterbath is heated rapidly to 60 °C and then slowly until the compound melts. This second trial should produce an accurate melting point with a 1–2 °C range; for example 69.5–71.0 °C.

EQUIPMENT and CHEMICALS

- ring stand & ring
- wire gauze
- mortar and pestle
- 110 °C thermometer with split cork
- 400-mL beaker
- 25 x 150 mm test tube containing 20 g of paradichlorobenzene
- 50 cm of 6-mm glass tubing
- capillary tubes
- rubber bands
- biphenyl (diphenyl)
- melting point unknowns

PROCEDURE

A. Cooling Curve and Freezing Point

The Instructor may wish to have students work in pairs. One student should set up the apparatus and record data while the other student heats the paradichlorobenzene and later observes temperature readings.

1. Set up the apparatus as shown in Figure 4.2. Add 300 mL of distilled water to the 400-mL beaker. Heat the water to 40 °C, and shut off the burner.

2. Obtain a test tube containing melted paradichlorobenzene at ~90 °C.

   Note: The Instructor will provide a large waterbath with several test tubes containing paradichlorobenzene liquid at ~90 °C.

3. Transfer the test tube and thermometer into the 400-mL beaker of water at 40 °C. Support the test tube with a utility clamp, and hold the thermometer using a split cork as shown in Figure 4.2.

4. Begin recording thermometer readings when the temperature drops to 65.0 °C. Continue recording the temperature (± 0.5 °C) every 30 seconds for ten minutes.

5. Plot the temperature/time data on the graph paper provided. Circle each point and draw a smooth cooling curve. Extend a dashed line from the flat portion of the curve to the vertical axis in order to determine the freezing point of the compound (see Figure 4.1).

   Note: If the thermometer is frozen in solid paradichlorobenzene, ***do not attempt to pull out the thermometer***. Return the test tube to the hot waterbath and allow the solid to melt; then remove the thermometer. Do not pour out the liquid paradichlorobenzene, as the compound is used for repeated trials.

B. Melting Point of an Unknown

1. Seal one end of a capillary tube with a burner flame. Let the tube cool, and then dab the open end into a small sample of biphenyl. Invert the capillary and lightly tap the sealed end to pack the sample. Repeat this process until a 5-mm sample is packed at the sealed end of the capillary.

   Note: If the biphenyl crystals are large, grind the crystals using a mortar and pestle. To pack the crystals, drop the sealed end of the capillary through a long piece of 6-mm glass tubing onto the lab bench.

2. Set up an apparatus as shown in Figure 4.3. Add 300 mL of distilled water into the 400-mL beaker. Attach the capillary at the end of the thermometer with a rubber band, and place in the beaker.

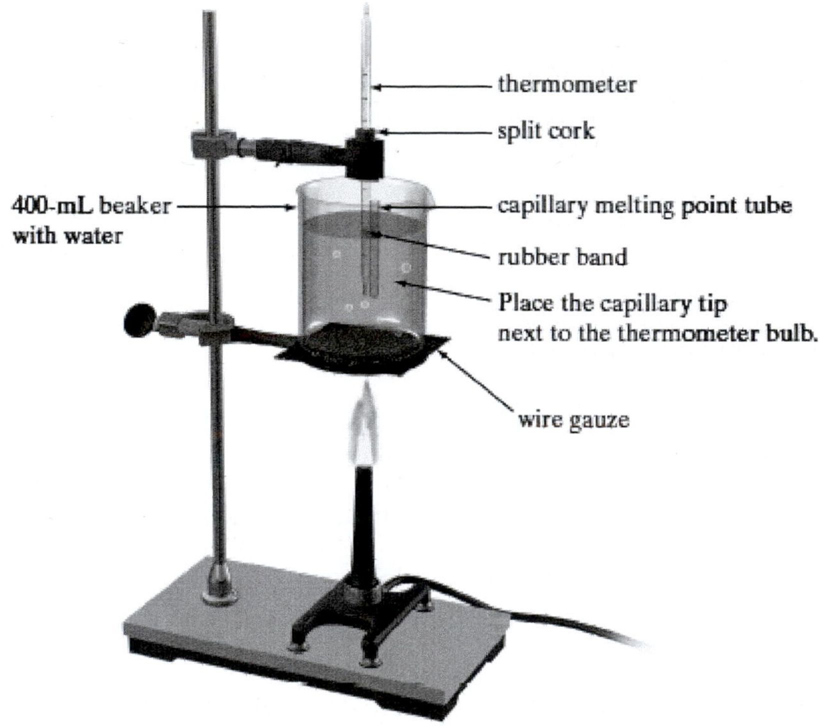

Figure 4.3 Melting Point Apparatus The melting point is recorded when the solid melts to a liquid and appears clear in the capillary tube.

3. Rapidly heat the water in the beaker until the biphenyl melts. Observe the approximate melting point (± 1 °C), and record the range of temperature in the Data Table.

4. Prepare another capillary tube and heat rapidly until the temperature is within 10 °C of the melting point. Then slowly continue to heat in order to determine the melting point accurately. Record the melting point range (± 0.5 °C) from the first sign of melting until the compound has completely melted. The reference value is given in the Data Table for comparison.

5. Obtain an unknown compound, and record the number. Determine the melting point for the unknown as above.

EXPERIMENT 4 NAME ______________________

DATE ______________ SECTION ______________________

*PRELABORATORY ASSIGNMENT**

1. Provide the key term that corresponds to each of the following definitions.

__________ (a) a term for the condition of a substance existing as a solid, liquid, or gas

__________ (b) the conversion from one physical state to another

__________ (c) the temperature at which a liquid substance crystallizes and forms a solid

__________ (d) the temperature at which a solid substance melts and forms a liquid

__________ (e) the horizontal axis (x-axis) on a graph

__________ (f) the vertical axis (y-axis) on a graph

__________ (g) the point of intersection of the horizontal and vertical axes on a graph

Key Terms: abscissa, change of state, freezing point, melting point, ordinate, origin, physical state

2. Why must distilled water be used in the hot waterbath?

3. When the test tube with hot liquid paradichlorobenzene is placed in the beaker of water to cool, what is the initial temperature of the water in the beaker?

4. When the test tube with hot liquid paradichlorobenzene is placed in the beaker of water, what is the initial temperature of the paradichlorobenzene?

5. What is the initial ***recorded*** temperature for the cooling curve data?

6. Which point on the cooling curve corresponds to the freezing point of paradichlorobenzene?

---

* *Answers in Appendix J*

7. After the liquid paradichlorobenzene freezes to a solid, how is the thermometer removed from the solid paradichlorobenzene?

8. Why are two trials performed to determine an accurate melting point of a compound?

9. While performing the first melting point trial, a compound begins to melt at 75 °C and liquefies completely at 85 °C. Report the approximate melting point.

10. While performing the second melting point trial, a compound begins to melt at 79.0 °C and liquefies completely at 80.5 °C. Report the precise melting point.

11. A solid compound in a capillary tube is placed in the waterbath and appears to liquefy before heating. Give two possible explanations for the observation.

    (1)

    (2)

12. What safety precautions must be observed in this experiment?

EXPERIMENT 4 NAME ____________________

DATE ____________ SECTION ____________________

DATA TABLE

A. Cooling Curve and Freezing Point

Temperature	*Time*	*Observation*
65.0 °C	0:00	liquid
	0:30	
	1:00	
	1:30	
	2:00	
	2:30	
	3:00	
	3:30	
	4:00	
	4:30	
	5:00	
	5:30	
	6:00	
	6:30	
	7:00	
	7:30	
	8:00	
	8:30	
	9:00	
	9:30	
	10:00	

B. Melting Point of an Unknown

	Rapid Trial	*Trial 2*
Mp of biphenyl (69–71 °C)	________ °C	________ °C
Mp of **UNKNOWN** # ______	________ °C	________ °C

RECYCLE Chemical Waste

Cooling Curve — Trial 2

Temperature	*Time*	*Observation*
65.0 °C	0:00	liquid
	0:30	
	1:00	
	1:30	
	2:00	
	2:30	
	3:00	
	3:30	
	4:00	
	4:30	
	5:00	
	5:30	
	6:00	
	6:30	
	7:00	
	7:30	
	8:00	
	8:30	
	9:00	
	9:30	
	10:00	

A. Cooling Curve — Trial 1 Freezing Point:________°C

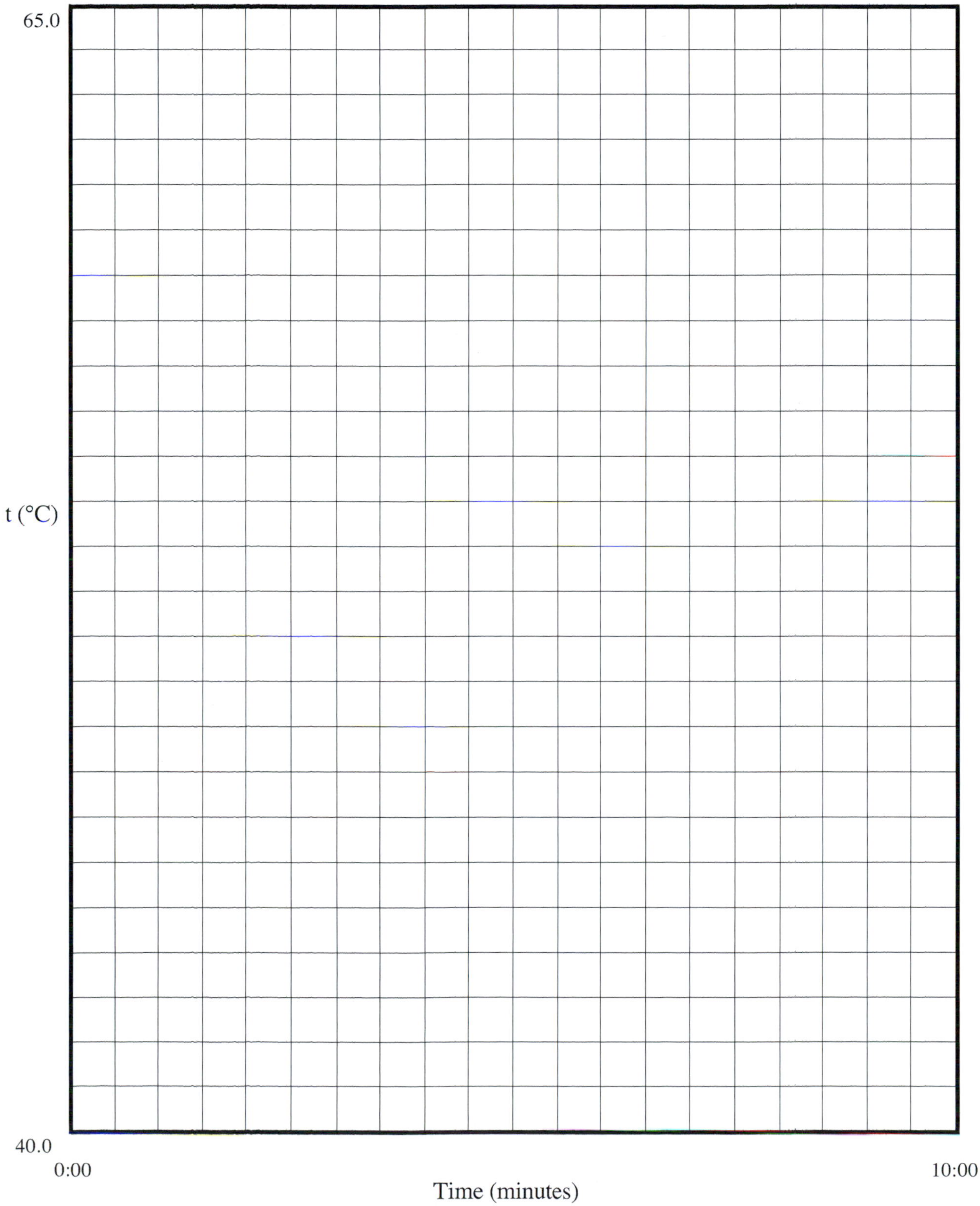

A. Cooling Curve — Trial 2 Freezing Point:________°C

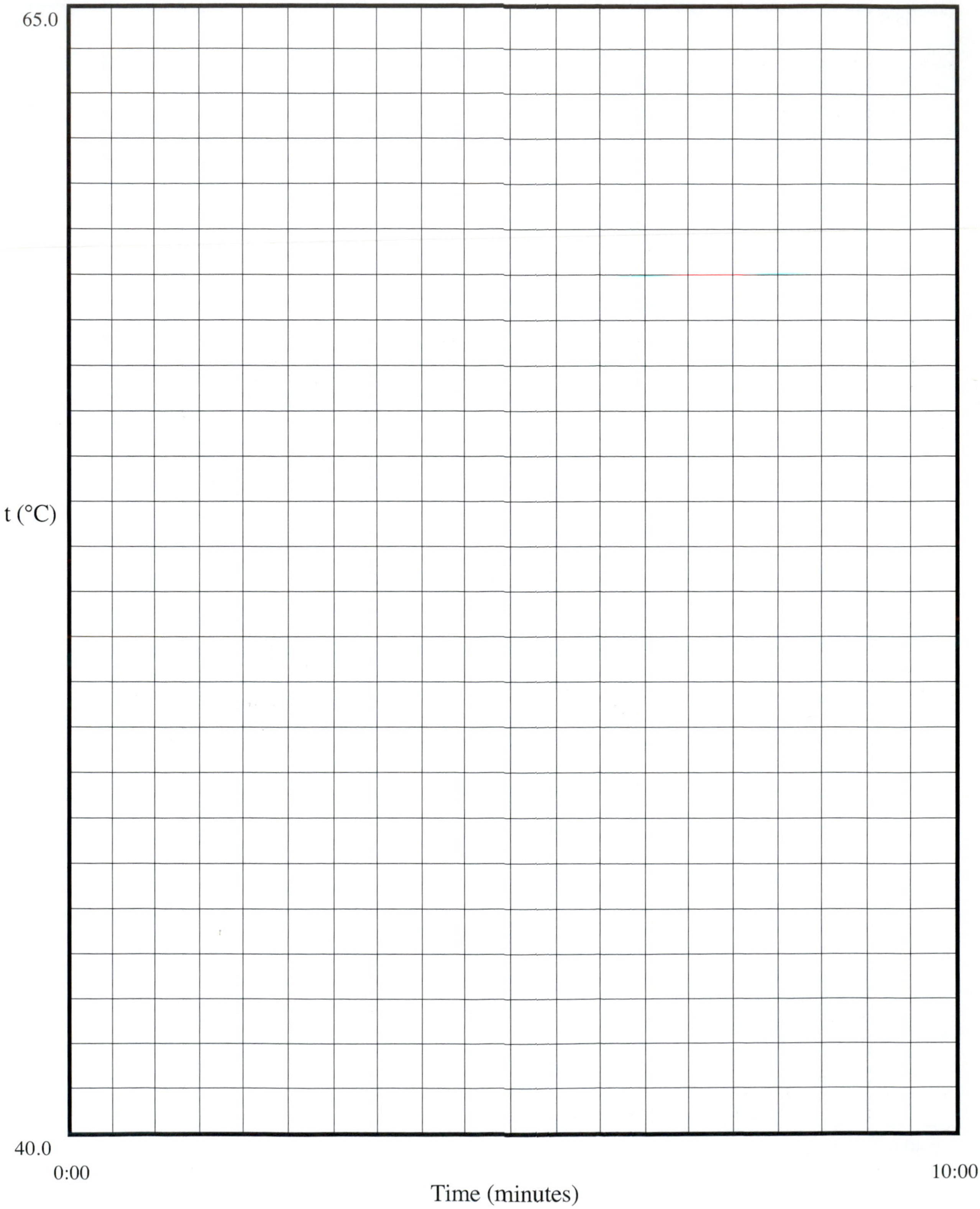

EXPERIMENT 4 NAME ______________________

DATE ______________ SECTION ______________________

POSTLABORATORY ASSIGNMENT

1. Naphthalene is a compound used in closets to destroy moth larva and protect clothes. Use the following data to graph the cooling curve for naphthalene.

Temperature (°C)	*Time (minutes)*
83.0	0:00
81.5	0:30
81.0	1:00
80.5	1:30
80.5	2:00
80.5	2:30
80.5	3:00
80.5	3:30
80.5	4:00
80.0	4:30
79.5	5:00

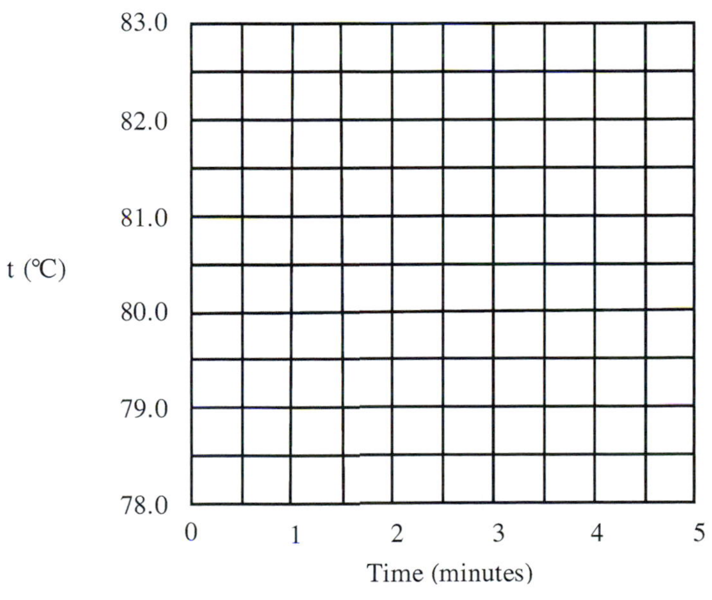

From the graph, estimate the freezing point of naphthalene (± 0.5 °C). ______________

2. Antifreeze freezes to a solid at 261 K. Calculate the freezing point of antifreeze on the Celsius and Fahrenheit temperature scales.

______________ °C

______________ °F

3. The following graph shows a heating curve for an unknown compound:

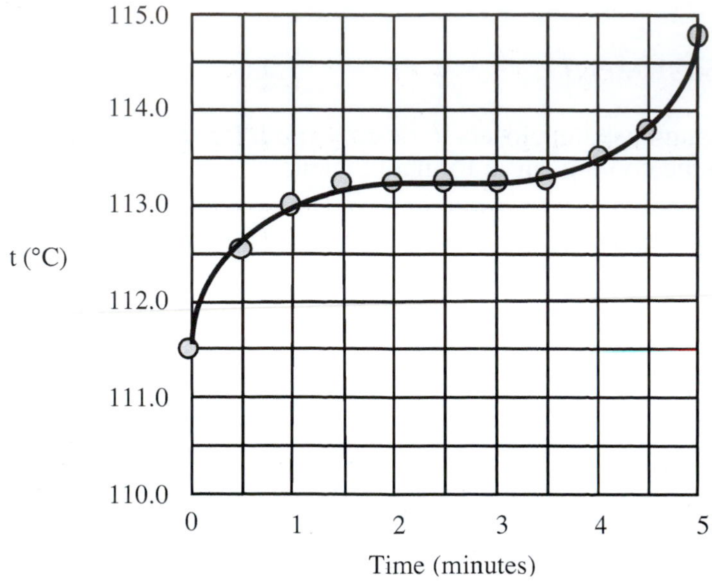

From the graph, estimate the melting point of the unknown (± 0.5 °C). ________________

4. What is the term for chunks of dry ice, solid CO_2, changing directly from a white solid to a gas at –78.5 °C?

____

5. What is the term for water vapor, gaseous H_2O, changing directly from a colorless gas to a white solid in a freezer at 0.0 °C?

____

6. (optional) Refer to the *Handbook of Chemistry and Physics, Physical Constants of Inorganic Compounds*, and find the melting points (°C) of the following elements.

   (a) gold, Au ________________

   (b) gallium, Ga ________________

EXPERIMENT

5

Physical Properties and Chemical Properties

OBJECTIVES

- To observe a demonstration of oxidation of a metal.
- To observe a demonstration of sublimation and deposition.
- To observe the appearance of several elements.
- To determine the boiling points of methyl alcohol and an unknown liquid.
- To determine whether a solid is *soluble* or *insoluble* in water.
- To determine whether a liquid is *soluble* or *insoluble* in water.
- To determine whether a substance is undergoing a *physical* or *chemical change*.
- To gain proficiency in determining a boiling point.

DISCUSSION

Chemists classify matter according to its physical and chemical properties. Matter can be classified as a mixture or a pure substance, depending upon its properties. A **heterogeneous mixture** has physical and chemical properties that vary within the sample. For example, combining sugar and salt gives a heterogeneous mixture because the properties of sugar and salt are different.

A **homogeneous mixture** has constant properties although the properties can vary from sample to sample. A homogeneous mixture may be a gaseous mixture, a solution, or an alloy. Examples include air, seawater, and brass, which is an alloy of the metals copper and zinc.

A pure **substance** is either a **compound** or an **element**. A pure substance has constant and predictable properties; examples include sodium chloride (compound), as well as sodium metal and chlorine gas (elements).

Water is a compound containing the elements hydrogen and oxygen. When electricity is passed through water, it decomposes into hydrogen gas and oxygen gas. Although hydrogen and oxygen are both colorless, odorless gases, they differ in their other physical and chemical properties. Figure 5.1 illustrates the overall relationship for the classification of matter.

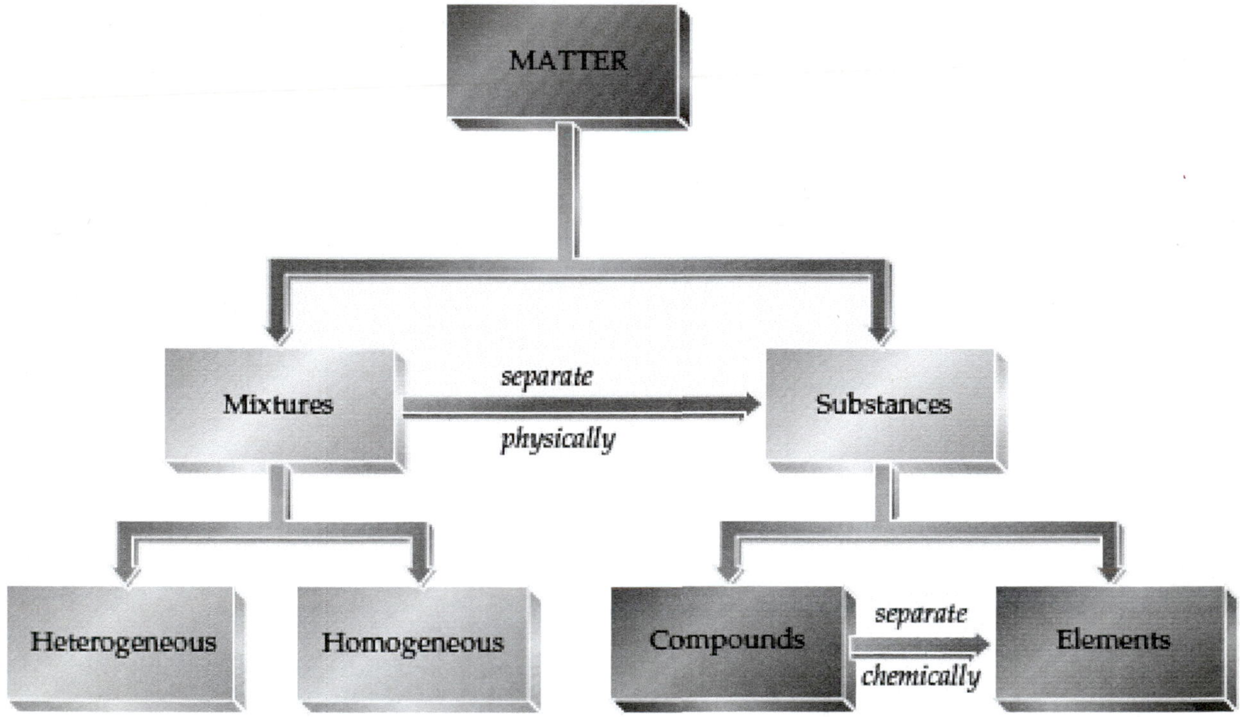

Figure 5.1 Classification of Matter Matter is classified as either a mixture or a pure substance. The properties of a *heterogeneous mixture* vary within the sample, but the properties of a *homogeneous mixture* are constant. A pure substance is either a *compound* or an *element*.

A **physical property** refers to a characteristic that can be observed without changing the composition of the substance. A partial list of important physical properties include: appearance, physical state (solid, liquid, or gas), density, malleability, ductility, conductivity of heat and electricity, melting point, boiling point, and solubility in water. A **chemical property** refers to a property that can only be observed during a chemical reaction. The chemical properties of oxygen gas include its ability to react with most metals and nonmetals. On the other hand, helium is an inert gas and does not react with other elements.

In this experiment, we will observe a **physical change** as a substance undergoes a change in physical state, a temporary change in color, or a simple change in volume when two solutions are added together. We will observe a **chemical change** when a substance releases a gas, undergoes a permanent change in color, or forms an insoluble substance when two solutions are added together. An antacid tablet fizzing in water, a banana changing color from green to yellow, and the formation of an insoluble "bathtub ring," are all familiar and practical examples of a chemical change.

EQUIPMENT and CHEMICALS

- ring stand & ring
- wire gauze
- hotplate (optional)
- 400-mL beaker
- 16 x 150 mm test tube
- boiling chip
- 110 °C thermometer
- split cork
- 13 x 100 mm test tubes (6) & test tube rack
- test tube holder
- test tube brush
- wash bottle with distilled water

- copper wire, heavy gauge Cu
- iodine, solid crystals I_2
- small vials with samples of cobalt, hydrogen, magnesium, manganese, neon, oxygen, silicon, sulfur, tin, zinc
- methyl alcohol, CH_3OH
- boiling point unknowns
- copper(II) sulfate, $CuSO_4 \cdot 5H_2O$
- calcium carbonate crystals, $CaCO_3$
- amyl alcohol (pentanol), $C_5H_{11}OH$
- ammonium bicarbonate, solid NH_4HCO_3
- potassium bicarbonate, solid $KHCO_3$
- sodium carbonate solution, 0.5 M Na_2CO_3
- sodium sulfate solution, 0.1 M Na_2SO_4
- dilute hydrochloric acid, 6 M HCl
- calcium nitrate solution, 0.1 M $Ca(NO_3)_2$
- copper(II) nitrate solution, 0.1 M $Cu(NO_3)_2$
- ammonium hydroxide solution, 6 M NH_4OH

PROCEDURE

A. Instructor Demonstrations

1. *Heating Copper Wire*

Show a piece of copper wire to the class. Hold the wire with crucible tongs and heat until the wire is red hot. Allow the wire to cool and show the class the wire after heating. Classify the observation as a *physical change* or *chemical change*.

RECYCLE Chemical Waste

Note: If students suggest the copper wire is covered with carbon from the burner flame (i.e., a physical change), the instructor can quickly disprove this by placing a piece of charcoal in one test tube and the copper wire in a second test tube. After adding hydrochloric acid to each test tube, students can observe a change in one test tube and no change in the other.

2. *Heating Iodine Crystals*

Put 3 small crystals of iodine in a dry 250-mL beaker. Cover the beaker using an evaporating dish containing ice. Support the beaker on a ring stand (see Figure 5.2) and gently heat the crystals until all the iodine vaporizes. Using crucible tongs to hold the hot evaporating dish, show the class the bottom of the evaporating dish and classify the observation as a *physical change* or *chemical change*.

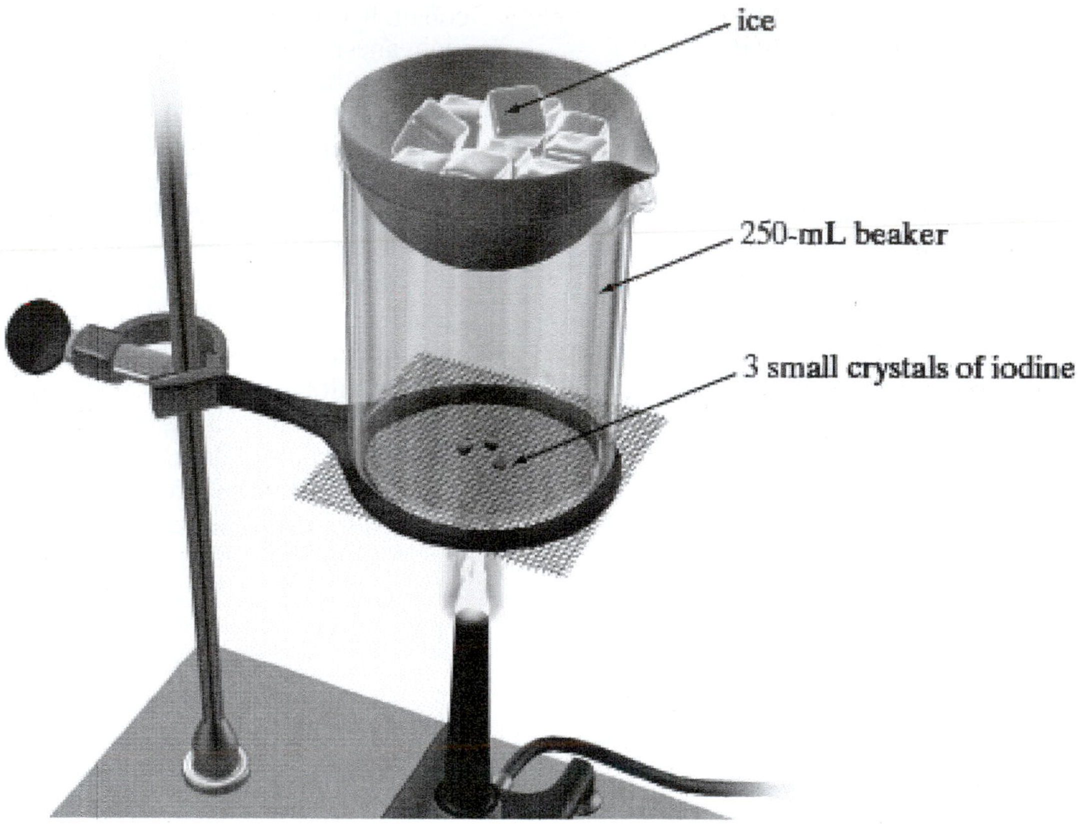

Figure 5.2 Apparatus for Sublimation/Deposition Gently heat a few crystals of iodine in the beaker. The iodine crystals undergo *sublimation*, which in turn undergoes *deposition* on the bottom of the evaporating dish.

B. Observation of Elements

Observe vials of the following elements and record your observations in the Data Table. Classify each element as a metal, semimetal, or nonmetal.

(a) cobalt
(b) hydrogen
(c) magnesium
(d) manganese
(e) neon
(f) oxygen
(g) silicon
(h) sulfur
(i) tin
(j) zinc

C. Physical Properties

1. *Boiling Point*
   (a) Support a 400-mL beaker on a ring stand as shown in Figure 5.3. Add 300 mL of distilled water to the beaker, bring to a boil, and shut off the burner. Add a boiling chip and 20 drops of methyl alcohol into a 16 x 150 mm test tube. Place the test tube in the hot water and suspend a thermometer about 1 cm above the liquid.

   (b) After the alcohol begins to boil in the test tube, record the boiling point (± 0.5°C) when alcohol drips from the tip of the thermometer every few seconds.

   Caution: Methyl alcohol is flammable; keep away from flames.

   (c) Record the number of an unknown liquid, and determine the boiling point of the liquid (± 0.5°C) as above.

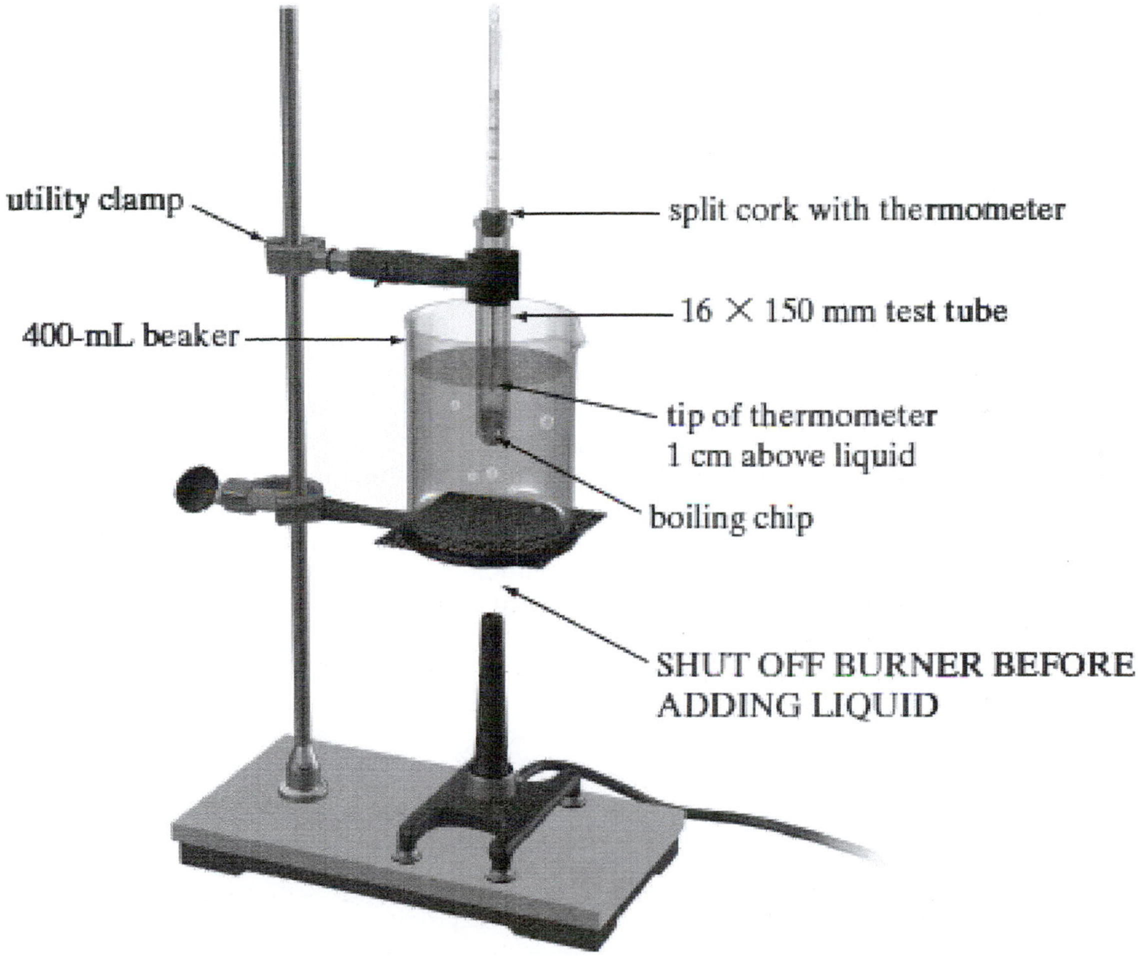

Figure 5.3 Boiling Point Apparatus The boiling point is recorded as vapor condenses on the tip of the thermometer and drips every 2 or 3 seconds.

Alternate Apparatus: If hotplates are available, the Instructor may wish to use a hotplate rather than the laboratory burner. Using a hotplate, bring the water in the beaker to a boil, and turn off the hotplate.

2. *Solubility of a Solid in Water*

Add 20 drops of distilled water into two test tubes. Drop a copper sulfate crystal into one test tube, and a calcium carbonate crystal into the other. Shake the test tubes briefly to observe solubility. State whether each solid is *soluble* or *insoluble* in water.

3. *Solubility of a Liquid in Water*

Add 20 drops of distilled water in two test tubes. Add a few drops of methyl alcohol to one test tube, and amyl alcohol to the other. Shake the test tubes briefly to mix the liquids. State whether each liquid is *soluble* or *insoluble* in water.

D. Chemical Properties

1. *Reactions of Compounds*

RECYCLE
Chemical
Waste

(a) Put a pea-sized sample of ammonium bicarbonate into a small test tube. Use a test tube holder and heat gently with a cool flame and note any changes, including odor. Classify your observation as a *physical change* or a *chemical change*.

(b) Put a pea-sized sample of potassium bicarbonate into a small test tube. Use a test tube holder and heat gently with a cool flame and record any changes. Classify your observation as a *physical change* or *chemical change*.

2. *Reactions of Solutions*

(a) Add 20 drops of sodium carbonate, and 20 drops of sodium sulfate into separate test tubes. Add 20 drops of dilute hydrochloric acid to each test tube, and record any changes. Classify your observation as a *physical change* or *chemical change*.

(b) Add 20 drops of calcium nitrate, and 20 drops of copper(II) nitrate into separate test tubes. Add 20 drops of dilute ammonium hydroxide to each test tube, and note any changes. Classify your observation as a *physical change* or *chemical change*.

EXPERIMENT 5 NAME ______________________

DATE ______________ SECTION ______________________

PRELABORATORY ASSIGNMENT*

1. Provide the key term that corresponds to each of the following definitions.

_________ (a) matter having an indefinite composition and properties that can vary within the sample

_________ (b) matter having a definite composition but properties that can vary from sample to sample; examples include alloys and solutions

_________ (c) matter having constant composition with definite and predictable properties

_________ (d) a pure substance that can be broken down into two or more simpler substances by chemical reaction

_________ (e) a pure substance that cannot be broken down any further by chemical reaction

_________ (f) a characteristic of the substance that can be observed without changing its chemical formula

_________ (g) a characteristic of a substance that cannot be observed without changing its chemical formula

_________ (h) a modification of a substance that does not alter its chemical composition

_________ (i) a modification of a substance that alters its chemical composition

_________ (j) an insoluble solid substance produced from a reaction in aqueous solution

Key Terms: chemical change, chemical property, compound, element, heterogeneous mixture, homogeneous mixture, physical change, physical property, precipitate, substance

2. Classify the following characteristics as a physical (*phys*) or chemical (*chem*) property.
   (a) physical state
   (b) density
   (c) melting point
   (d) hardness
   (e) appearance
   (f) reactivity
   (g) solubility
   (h) conductivity

3. Classify the following observations as a physical (*phys*) or chemical (*chem*) change.
   (a) candle burning
   (b) wax melting
   (c) alcohol vaporizing
   (d) antacid fizzing in water
   (e) apple turning brown
   (f) steam condensing on a mirror
   (g) fire releasing heat
   (h) fireworks releasing light

---

* *Answers in Appendix J*

4. What is the purpose of the boiling chip when determining the boiling point of a liquid?

5. What experimental observations indicate a chemical change is taking place?

6. What experimental observations indicate a gas is being released?

7. What safety precautions must be observed in this experiment?

EXPERIMENT 5 NAME ______________________

DATE ______________ SECTION ______________________

DATA TABLE

A. Instructor Demonstrations

Procedure	*Observation*	*Physical Change or Chemical Change*
1. *Heating Copper Wire*		
2. *Heating Iodine Crystals*		

RECYCLE Chemical Waste

B. Observation of Elements

Element	*Symbol*	*Physical State*	*Color*	*Metal/Semimetal/ or Nonmetal*
cobalt				
hydrogen				
magnesium				
manganese				
neon				
oxygen				
silicon				
sulfur				
tin				
zinc				

C. Physical Properties

1. *Boiling Point*

Bp of methyl alcohol (65.0°C) ________________ °C

Bp of **UNKNOWN** #______ ________________ °C

2. *Solubility of a Solid in Water*

copper sulfate crystal and water ________________

calcium carbonate crystal and water ________________

3. *Solubility of a Liquid in Water*

methyl alcohol (methanol) and water ________________

amyl alcohol (pentanol) and water ________________

D. Chemical Properties

Procedure	*Observation*	*Physical Change or Chemical Change*
1. *Reactions of Compounds*		
(a) ammonium bicarbonate + heat		
(b) potassium bicarbonate + heat		
2. *Reactions of Solutions*		
(a) sodium carbonate + hydrochloric acid		
sodium sulfate + hydrochloric acid		
(b) calcium nitrate + ammonium hydroxide		
copper(II) nitrate + ammonium hydroxide		

RECYCLE Chemical Waste

EXPERIMENT 5 NAME ____________________

DATE ______________ SECTION ____________________

POSTLABORATORY ASSIGNMENT

1. State whether the following properties are more typical of a *metal* or a *nonmetal* element.

   (a) silver solid ________ (b) yellow powder ________

   (c) ductile solid ________ (d) colorless gas ________

   (e) high melting point ________ (f) poor conductor ________

2. Classify each of the following as an example of an *element*, *compound*, *homogeneous mixture*, or *heterogeneous mixture*.

   (a) copper, Cu ________ (b) copper alloy ________

   (c) copper oxide, CuO ________ (d) copper ore ________

3. Classify the following as a physical property (*phys*) or a chemical property (*chem*).

   (a) Copper metal has a red-orange metallic luster. ________

   (b) Copper metal has a density of 8.94 g/cm^3. ________

   (c) Copper metal and chlorine gas produce $CuCl_2$. ________

   (d) Copper metal has a melting point of 1084 °C. ________

   (e) Copper metal conducts electricity. ________

   (f) Copper metal and acid give no reaction. ________

4. Classify the following as a physical change (*phys*) or a chemical change (*chem*).

   (a) Silver tarnishes when exposed to air. ________

   (b) Baking soda fizzes when added to vinegar. ________

   (c) Alcohol dissolves when added to water. ________

   (d) White phosphorus glows when exposed to air. ________

   (e) Soap and tap water form an insoluble "bathtub ring," ________

   (f) Water evaporates from a lake. ________

5. Refer to the *Handbook of Chemistry and Physics*, *The Elements*, to research the density, melting point, and boiling point for the following elements.

Name	*density* *($g \cdot cm^{-3}$)*	*mp* *(°C)*	*bp* *(°C)*
Cobalt	________	________	________
Magnesium	________	________	________
Manganese	________	________	________
Silicon	________	________	________
Zinc	________	________	________

6. (optional) Go online to Wikipedia at www.wikipedia.org to research the density, melting point, and boiling point for the following elements.

Name	*density* *($g \cdot cm^{-3}$)*	*mp* *(°C)*	*bp* *(°C)*
Cobalt	________	________	________
Magnesium	________	________	________
Manganese	________	________	________
Silicon	________	________	________
Zinc	________	________	________

EXPERIMENT

6

"Atomic Fingerprints"

OBJECTIVES

- To distinguish between a *continuous spectrum* and a *line spectrum.*
- To compare observed and calculated lines in the hydrogen spectrum.
- To identify unknown elements in a fluorescent light by "atomic fingerprints."
- To become proficient in using a hand spectroscope.

DISCUSSION

In 1900, Max Planck, a German professor of physics, proposed a revolutionary concept. The concept was quite simple: since matter consists of particles, energy may consist of particles. This concept was revolutionary because light energy had previously been considered a continuous wave of energy rather than a stream of particles. Scientists refer to a particle of matter as an atom, and refer to a particle of light energy as a quantum, or a **photon**.

Although the terms light and radiant energy are often used interchangeably, **light** usually refers to radiant energy that is visible, rather than invisible radiation such as infrared or ultraviolet. Light travels as a wave, and the **wavelength** is the crest-to-crest distance to complete one cycle. The **frequency** is the number of wave cycles that occur in one second. Low-energy light has a long wavelength and a low frequency. Conversely, high-energy light has a short wavelength and a high frequency.

Radiant Energy Spectrum

When we observe white light, we see the effect of several combined colors of light. When white light passes through a glass prism, however, it separates into six primary colors: red, orange, yellow, green, blue, and violet. This is illustrated in Figure 6.1.

A rainbow is a natural phenomenon that results from sunlight passing through raindrops, which act as miniature prisms to separate sunlight into various bands of color. We can record the wavelength of light in nanometers, where a **nanometer** (**nm**) is one-billionth of a meter. The range of visible light, violet to red, is usually considered 400–700 nm.

We see in Figure 6.2 that the **visible spectrum** is only a small portion of the radiant energy spectrum. The entire spectrum includes X ray, ultraviolet, visible, infrared, and microwave radiation. Our eyes can detect light only in the visible spectrum and not in other regions. That is, the wavelengths of ultraviolet light are too short to be seen by the human eye, and the wavelengths of infrared light are too long to be visible.

Bohr Model of the Atom

In 1913, Niels Bohr proposed that electrons travel in circular orbits around the nucleus, much as planets travel in orbits around the Sun. The Bohr model was a beautiful theory of electrons in atoms. However, no one knew if the model was right or wrong because there was no experimental evidence for the theory. Coincidentally, about this time, Bohr received an article showing the emission of light from hydrogen gas. Bohr saw that hydrogen gas emits colored lines, rather than a **continuous spectrum**. The three most prominent lines are violet, blue-green, and red.

If hydrogen gas in a glass tube is subjected to electricity, the excited hydrogen gas releases energy as reddish-purple light. When the emitted light passes through a glass prism, the reddish-purple glow separates into narrow bands of light. This collection of narrow bands of light is called a **line spectrum**. Figure 6.3 illustrates the line spectrum of hydrogen.

As Bohr considered the emission spectrum of hydrogen, he realized that he had powerful experimental evidence for his model of the atom. His concept of energy levels was supported by the line spectrum of hydrogen. He reasoned as follows: When hydrogen is energized using a gas discharge tube, electrons in hydrogen atoms jump to a higher energy level. For example, electrons may jump from the first level to the second, third, fourth, or fifth level. Next, excited electrons lose energy by dropping to a lower level closer to the nucleus.

When an electron drops to a lower energy level, it loses a specific amount of energy that corresponds to the energy of one photon of light. That is, the emitted photon of light has the same amount of energy as that lost by the electron when it drops from a higher to lower energy level. Thus, the line spectrum of hydrogen gas supports Bohr's model of the atom *experimentally*. However, it was Planck's idea that light is composed of particles that supported Bohr's model of the atom *theoretically*. Figure 6.4 shows the line spectrum of hydrogen and the corresponding electron energy levels.

Further study of emission spectra revealed that each element produces a unique set of spectral lines. For this reason, the line spectrum of a given element is sometimes referred to as an "**atomic fingerprint**."

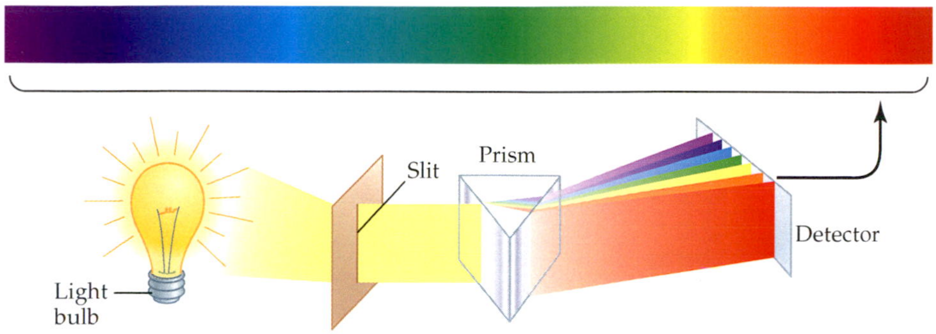

Figure 6.1 White Light Passing Through a Prism White light produces a rainbow when it passes through a glass prism. Similarly, sunlight produces a rainbow when it passes though raindrops, which act as miniature prisms.

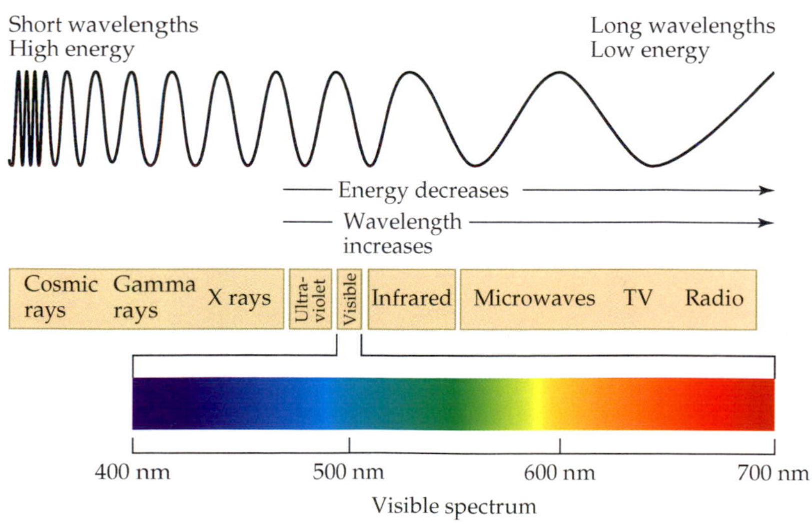

Figure 6.2 The Radiant Energy Spectrum The radiant energy spectrum includes short-wavelength cosmic rays through long-wavelength radio waves. The *ultraviolet* spectrum is approximately 100–400 nm, the *visible* spectrum is 400–700 nm, and the *infrared* region is 700–5000 nm.

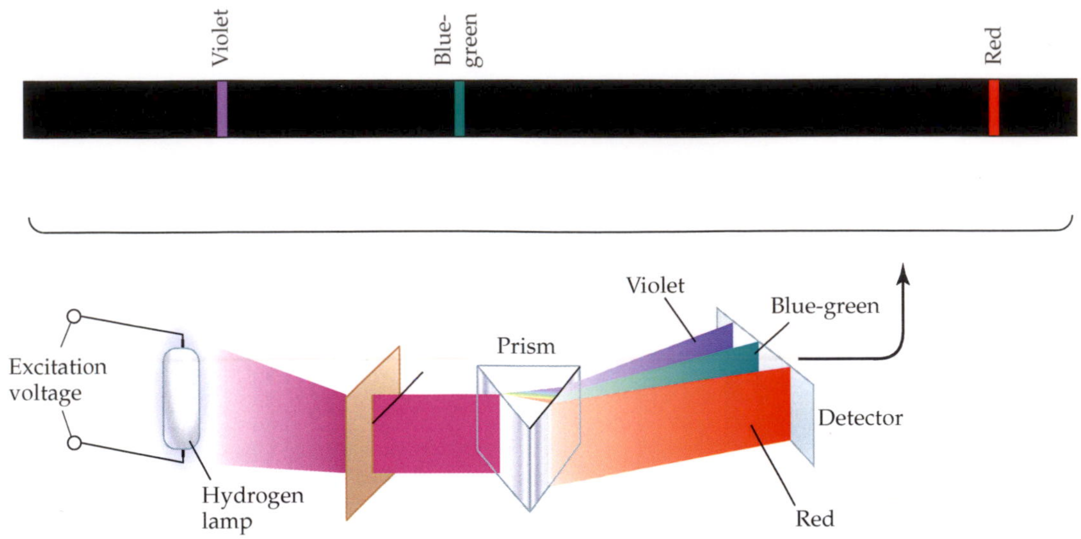

Figure 6.3 Hydrogen Line Spectrum A reddish-purple light is emitted when a voltage is applied to a hydrogen lamp. When the reddish-purple light passes through a glass prism, three lines are observed: violet, blue-green, and red.

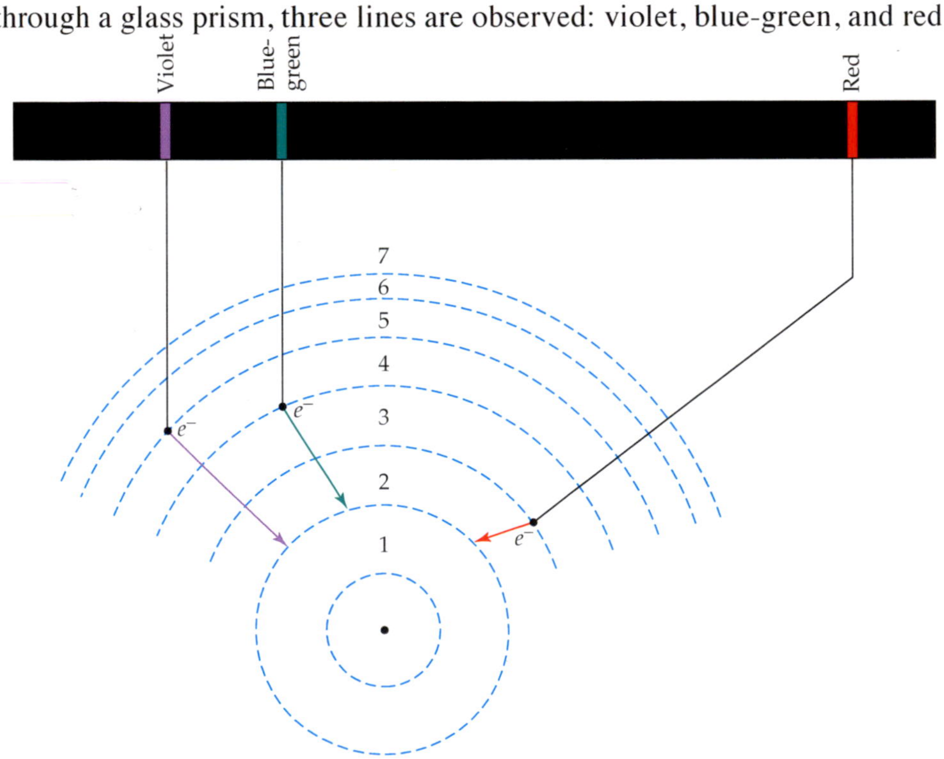

Figure 6.4 Hydrogen Spectral Lines and Energy Levels When electrons drop from energy level **5** to **2**, we observe a violet line. When electrons drop from level **4** to **2**, we see a blue-green line; from level **3** to **2**, we observe a red line.

Balmer Formula

In 1885, Johann Jakob Balmer, a Swiss mathematician and physicist, published a formula that accounts for the visible lines emitted from excited hydrogen gas. The **Balmer formula** shows that the wavelength (λ) of light for each line in the hydrogen spectrum is related to a small whole number (n) in the following way.

$$\frac{1}{\lambda} = \frac{1}{91\ \text{nm}}\left(\frac{1}{2^2} - \frac{1}{n^2}\right)$$

When Balmer set $n = 3$ in the formula, the calculated wavelength is that of the red line in the hydrogen spectrum. Similarly, substituting $n = 4$ and $n = 5$ gives values for the blue-green and violet lines. When Niels Bohr read Balmer's paper, he realized that the n value represents an energy level; and 2 in the formula corresponds to energy level 2.

The following example calculation illustrates the Balmer formula for an electron dropping from the sixth energy level to the second energy level in a hydrogen atom.

Example Exercise 6.1 • Balmer Formula

Calculate the wavelength of light corresponding to the energy released when an electron drops from $n = 10$ to $n = 2$ in a hydrogen atom.

Solution: We can use the Balmer formula to calculate the wavelength of the spectral line as follows.

$$\frac{1}{\lambda} = \frac{1}{91\ \text{nm}}\left(\frac{1}{2^2} - \frac{1}{n^2}\right)$$

$$\frac{1}{\lambda} = \frac{1}{91\ \text{nm}}\left(\frac{1}{2^2} - \frac{1}{10^2}\right)$$

$$\frac{1}{\lambda} = \frac{1}{91\ \text{nm}}\left(\frac{1}{4} - \frac{1}{100}\right)$$

$$\frac{1}{\lambda} = \frac{1}{91\ \text{nm}}(0.25 - 0.01)$$

$$\frac{1}{\lambda} = \frac{0.24}{91\ \text{nm}}$$

After taking the reciprocal, we obtain

$$\lambda = \frac{91\ \text{nm}}{0.24} = 379.17 = 380\ \text{nm}$$

Since 91 nm has two significant digits, the calculated value rounds to two significant digits; thus, 380 nm.

Rydberg Equation

In 1890, the Swedish physicist Johannes Rydberg derived a general formula for calculating the wavelengths of spectral lines from excited hydrogen gas. Rydberg was able to account for an electron dropping from any higher energy level (n_H) to any lower energy level (n_L). The Rydberg equation can be expressed as

$$\frac{1}{\lambda} = \frac{1}{91\ \text{nm}} \left(\frac{1}{n_L^2} - \frac{1}{n_H^2}\right)$$

The Balmer formula is adequate to explain the visible lines in the spectrum of hydrogen because excited electrons are dropping to $n = 2$. However, the Rydberg equation also accounts for electrons dropping to any energy level including $n = 1$. For example, when electrons drop from $n = 4$ to $n = 1$, the spectral line is in the ultraviolet region, which is not visible to the human eye. Similarly, when electrons drop from $n = 4$ to $n = 3$, the spectral line is in the infrared region of the spectrum, and is therefore not visible.

The following example calculation illustrates the Rydberg equation for an electron dropping from the fourth to the first energy level in a hydrogen atom.

Example Exercise 6.2 • Rydberg Equation

Calculate the wavelength of light released when an electron drops from $n = 4$ to $n = 1$ in a hydrogen atom.

Solution: We can use the Rydberg equation to find the wavelength of the line as follows.

$$\frac{1}{\lambda} = \frac{1}{91\ \text{nm}} \left(\frac{1}{n_L^2} - \frac{1}{n_H^2}\right)$$

$$\frac{1}{\lambda} = \frac{1}{91\ \text{nm}} \left(\frac{1}{1^2} - \frac{1}{4^2}\right)$$

$$\frac{1}{\lambda} = \frac{1}{91\ \text{nm}} \left(\frac{1}{1} - \frac{1}{16}\right)$$

$$\frac{1}{\lambda} = \frac{1}{91\ \text{nm}} (1 - 0.06)$$

$$\frac{1}{\lambda} = \frac{0.94}{91\ \text{nm}}$$

After taking the reciprocal and rounding to two significant digits, we obtain

$$\lambda = \frac{91\ \text{nm}}{0.94} = 96.8 = 97\ \text{nm}$$

A wavelength of 97 nm is below the visible spectrum (400–700 nm). If we refer to Figure 6.2, we find the line is in the *ultraviolet* region of the spectrum.

EQUIPMENT and CHEMICALS

- Science Kit Hand Spectroscope
  The SK student spectroscope is available from Science Kit, Inc. @ 1-800-828-7777 (www.sciencekit.com).
- spectral tubes: hydrogen, helium, neon, argon, krypton, and mercury
- spectral tube power supply
  The spectrum tubes and power supply are readily available; for example, Sargent-Welch @ 1-800-SARGENT (www.sargentwelch.com), or VWR Scientific @ 1-800-932-5000 (www.vwr.com).
- (optional) colored pencils: violet, blue, green, yellow, orange, red

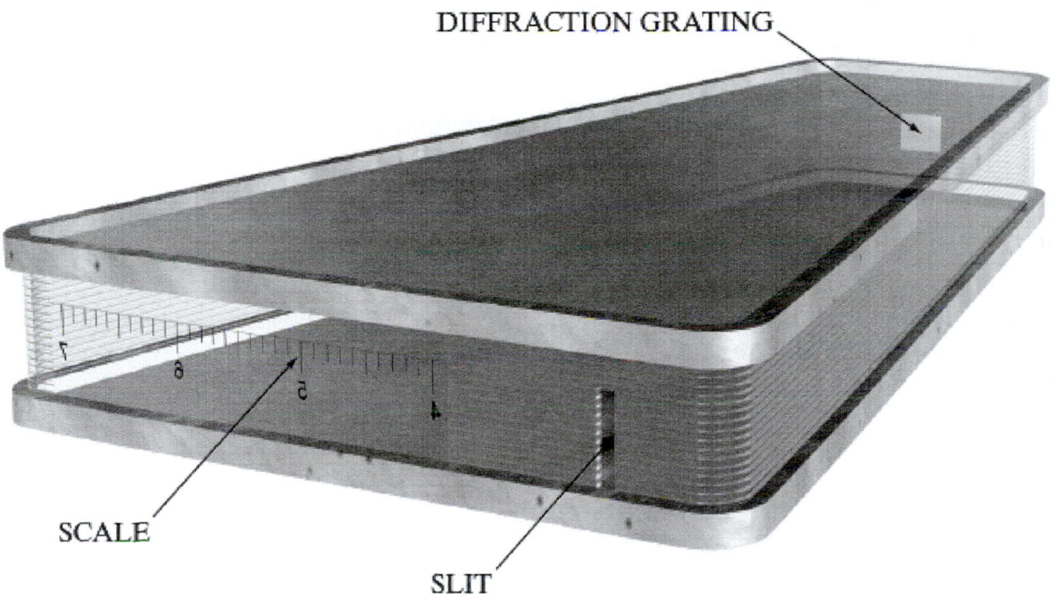

Figure 6.5 The Hand Spectroscope Light enters the spectroscope through the slit and strikes the diffraction grating. The grating is a thin piece of plastic with hundreds of parallel grooves that diffract light into different wavelengths.

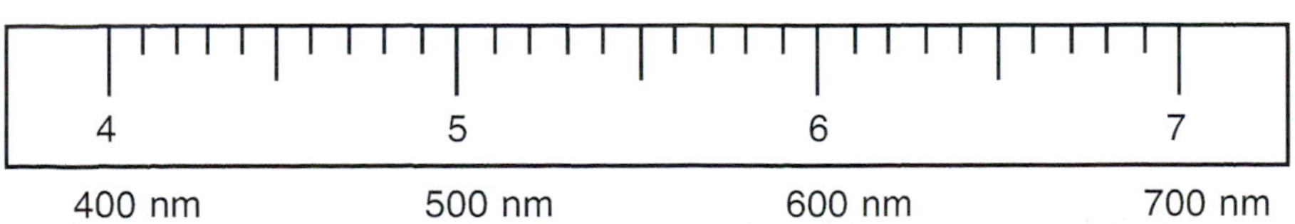

Figure 6.6 Wavelength Scale The spectroscope scale indicates wavelengths from 400 nm to 700 nm. The digit 4 is read as 400 nm, 5 as 500 nm, and 6 as 600 nm. There are ten subdivisions between each number on the scale; therefore each subdivision is 10 nm. For example, the scale divisions between 4 and 5 are read as 410, 420, 430, 440, 450, 460, 470, 480, and 490 nm.

PROCEDURE

A. Continuous Spectrum – White Light

1. With the hand spectroscope, observe the emission spectrum from one or more of the following: light from an overhead projector, light from an incandescent light bulb, and sunlight. Draw the observed spectrum on the wavelength scale in the Data Table.

    Note: It may be necessary to adjust the room lights in order to read the scale divisions in the hand spectroscope.

B. Line Spectrum – Hydrogen

1. Using a hand spectroscope, observe the hydrogen spectrum from a gas discharge tube. Draw the position of each line on the wavelength scale in the Data Table.

2. Using the Balmer formula, find the wavelength (λ) of light produced when electrons drop from energy level 3 to 2.

    Note: The calculated value should be rounded to two significant digits; for example, 655.2 nm rounds to 660 nm.

3. Repeat the wavelength calculation for the spectral lines produced when electrons drop from energy level 4 to 2; and from energy level 5 to 2.

4. Record the observed and calculated wavelength values in the Data Table. State the error after comparing the observed and calculated wavelengths.

C. Line Spectra – Helium, Neon, Argon, Krypton, and Mercury

1. Insert a helium gas discharge tube into a spectral tube power supply. Using a hand spectroscope, observe the emission lines from helium gas. Draw the position of six intense lines on the wavelength scale in the Data Table.

2. Repeat the procedure using a neon gas discharge tube, argon gas discharge tube, krypton gas discharge tube, and mercury vapor discharge tube in the power supply. Draw the position of the most intense lines on the wavelength scale in the Data Table.

D. Identifying Unknown Elements in a Fluorescent Light

1. Observe the line spectrum from a fluorescent light using the hand spectroscope. **Disregard the continuous rainbow**, and draw the position of each line on the wavelength scale in the Data Table. Compare the line spectrum from the fluorescent light to the lines in the emission spectra of He, Ne, Ar, Kr, and Hg. Identify the elements in the fluorescent light from their "atomic fingerprints."

    Note: Currently, there are two elements in a fluorescent light, one of which should be easy to identify based on its "atomic fingerprint." The second element will require a careful comparison of "fingerprints" to identify.

EXPERIMENT 6 NAME ______________________

DATE ______________ SECTION ______________________

PRELABORATORY ASSIGNMENT*

1. Provide the key term that corresponds to each of the following definitions.

__________ (a) a specific term that refers to the portion of the radiant energy spectrum that is visible

__________ (b) the region in the radiant energy spectrum from approximately 400–700 nm

__________ (c) the number of times a light wave travels a complete cycle in one second

__________ (d) the distance a light wave travels to complete one cycle

__________ (e) a particle of light that corresponds to a unit of radiant energy

__________ (f) a broad uninterrupted band of radiant energy

__________ (g) the unique line spectrum that is characteristic of a given element and can be used for identification

__________ (h) the narrow bands of light observed through a spectroscope that are emitted from excited atoms in a gas discharge tube

__________ (i) a mathematical formula for calculating the wavelength of light emitted from an excited hydrogen atom when the electron drops to the second energy level

__________ (j) a mathematical equation for calculating the wavelength of light emitted from an excited hydrogen atom when the electron drops to any lower energy level

Key Terms: "atomic fingerprint," Balmer formula, continuous spectrum, frequency, light, line spectrum, photon, Rydberg equation, visible spectrum, wavelength

2. Arrange the six primary colors (**R**, **O**, **Y**, **G**, **B**, **V**) in the visible spectrum in order of

(a) increasing wavelength

(b) increasing frequency

(c) increasing energy

* *Answers in Appendix J*

3. What wavelengths of light are indicated by the following three lines on the wavelength scale observed using a hand spectroscope?

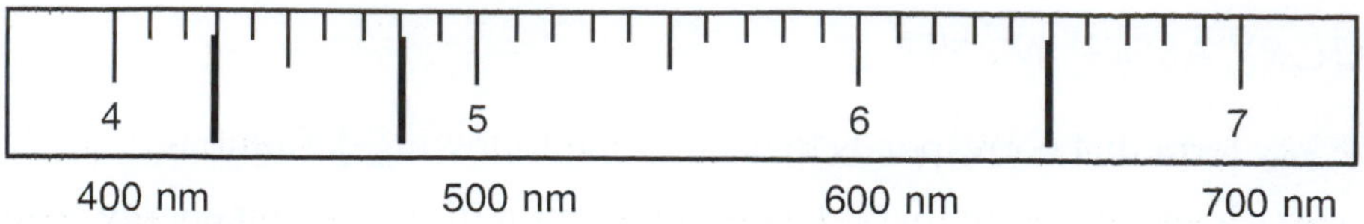

4. What are the colors of the three lines observed in the hydrogen line spectrum?

5. A faint line is observed when electrons in a hydrogen atom drop from energy level 6 to 2. Refer to the Balmer formula in Example Exercise 6.1 and calculate the wavelength for the line.

6. How many photons are emitted when 1 electron drops from energy level 6 to 2?

How many photons are emitted when 10 electrons drop from energy level 6 to 2?

7. What safety precautions must be observed in this experiment?

EXPERIMENT 6 NAME ______________________

DATE ______________ SECTION ______________________

DATA TABLE

A. Continuous Spectrum – White Light **Spectroscope #**__________

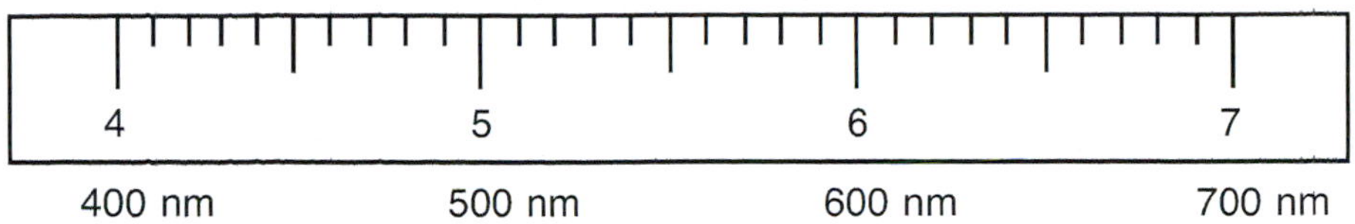

B. Line Spectrum – Hydrogen

1. Observed Wavelengths of Spectral Lines

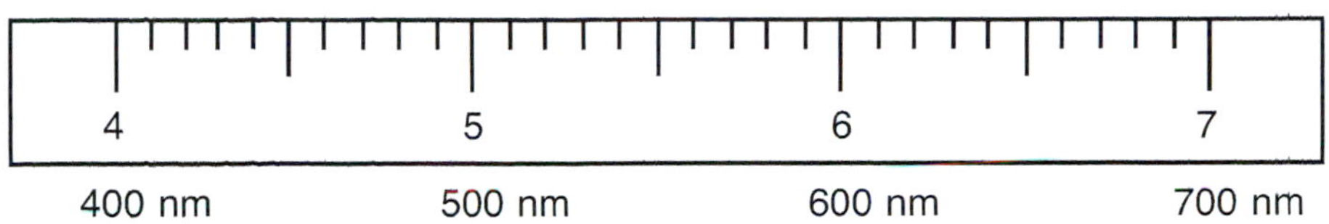

2. Calculated Wavelengths of Spectral Lines (see Example Exercise 6.1)

(a) red line (electrons drop from $n = 3$ to $n = 2$)

(b) blue-green line (electrons drop from $n = 4$ to $n = 2$)

(c) violet line (electrons drop from $n = 5$ to $n = 2$)

Spectral Line	*Observed Wavelength*	*Calculated Wavelength*	*Experimental Error*
red line	__________ nm	__________ nm	__________ nm
blue-green line	__________ nm	__________ nm	__________ nm
violet line	__________ nm	__________ nm	__________ nm

C. Line Spectra – Helium, Neon, Argon, Krypton, and Mercury

(a) Helium

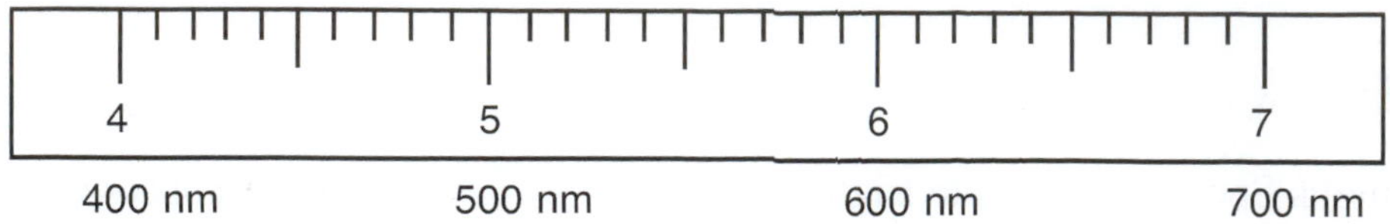

(b) Neon

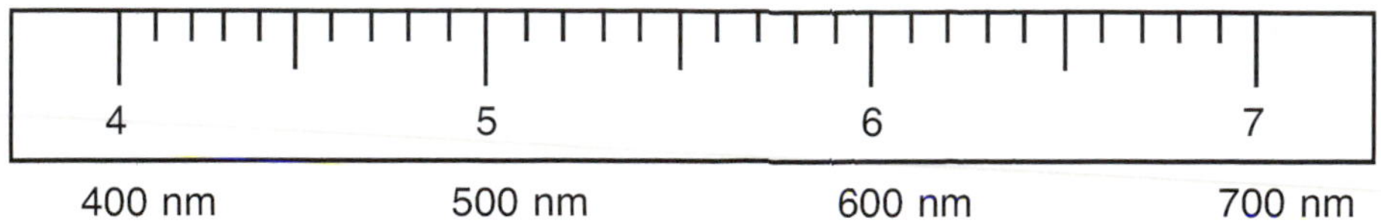

(c) Argon

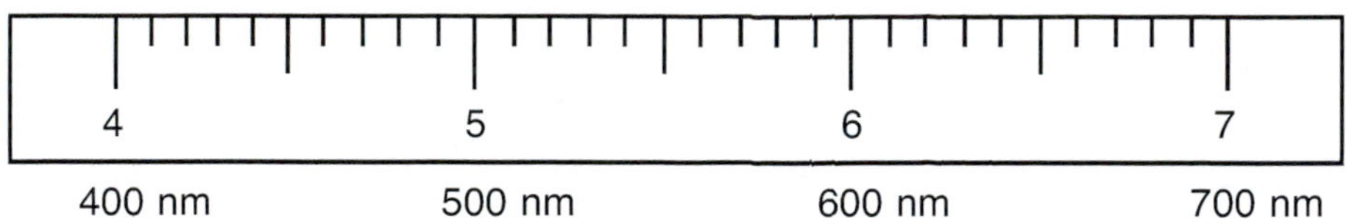

(d) Krypton

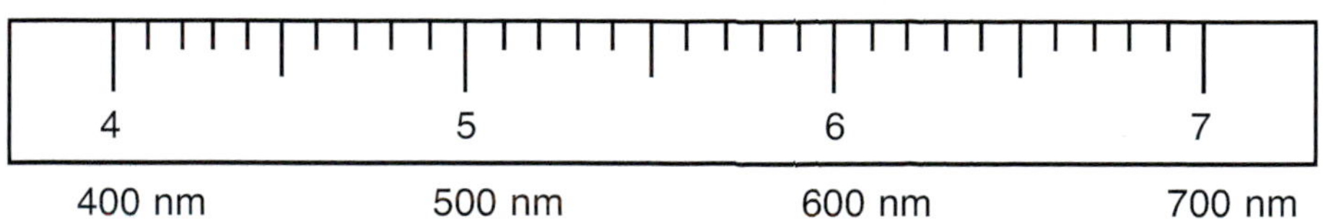

(e) Mercury

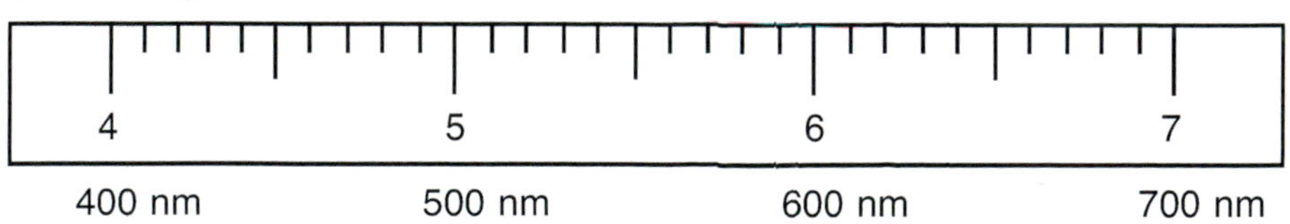

D. Identifying Unknown Elements in a Fluorescent Light

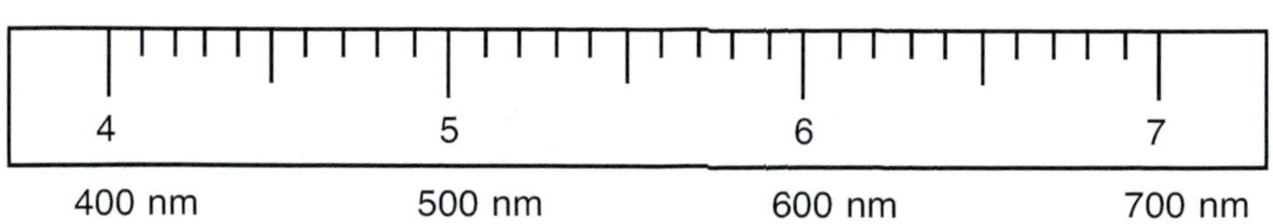

Based on its "atomic fingerprint," is **He** present in the fluorescent light? YES / NO

Based on its "atomic fingerprint," is **Ne** present in the fluorescent light? YES / NO

Based on its "atomic fingerprint," is **Ar** present in the fluorescent light? YES / NO

Based on its "atomic fingerprint," is **Kr** present in the fluorescent light? YES / NO

Based on its "atomic fingerprint," is **Hg** present in the fluorescent light? YES / NO

EXPERIMENT 6 NAME ____________________

DATE ____________________ SECTION ____________________

POSTLABORATORY ASSIGNMENT

1. What is the color of the hydrogen emission line corresponding to the following changes in energy level? (Refer to Figure 6.4.)

   (a) electrons drop from energy level 5 to level 2?

   (b) electrons drop from energy level 4 to level 2?

   (c) electrons drop from energy level 3 to level 2?

2. Arrange each of the following in order of *increasing* energy:

   (a) the changes of electron energy levels from 5 to 2, 4 to 2, and 3 to 2.

   (least energetic) ____to____ >>> ____to____ >>> ____to____ *(most energetic)*

   (b) the wavelengths of spectral lines 650 nm, 480 nm, and 430 nm.

   (least energetic) ________ >>> ________ >>> ________ *(most energetic)*

3. Calculate the wavelength of the emission line when electrons drop from energy level 2 to 1 in a hydrogen atom. (Refer to the Rydberg equation in Example Exercise 6.2.)

   Is this emission line in the *ultraviolet, visible,* or *infrared* region of the spectrum? (Refer to Figure 6.2.)

4. Calculate the wavelength of the emission line when electrons drop from energy level 4 to 3 in a hydrogen atom. (Refer to the Rydberg equation in Example Exercise 6.2.)

   Is this emission line in the *ultraviolet, visible,* or *infrared* region of the spectrum? (Refer to Figure 6.2.)

5. What region of the spectrum is next to ultraviolet, but is less energetic? (Refer to Figure 6.2.)

6. What region of the spectrum is next to infrared, but is more energetic? (Refer to Figure 6.2.)

7. How many photons of light are emitted when the electrons in a million hydrogen atoms drop from energy level 4 to 2?

8. What is the basic unit that represents each of the following?

(a) quantum nature of electricity

(b) quantum nature of light

9. State whether each of the following instruments gives a *continuous* or *quantized* measurement.

(a) metric ruler

(b) electronic balance

(c) graduated cylinder

(d) digital clock

10. Advertising lights are often referred to as "neon lights." If advertising lights are simply large gas discharge tubes, do all "neon lights" contain neon gas? Explain.

11. (optional) Go online to www.wikipedia.org to research the emission line spectrum for a "*mercury-vapor lamp*." Confirm the lines you observed for mercury vapor in Procedure C5.

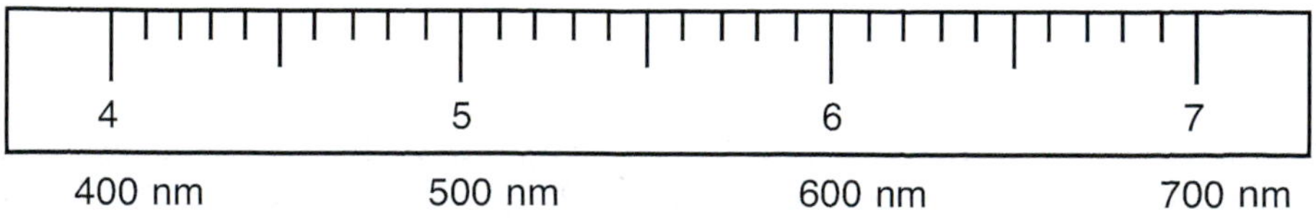

EXPERIMENT

7

Families of Elements

OBJECTIVES

- To study similar chemical properties for groups of elements in the periodic table.
- To observe flame tests and reactions for barium, calcium, lithium, potassium, sodium, and strontium solutions.
- To observe halide tests for bromide, chloride, and iodide solutions.
- To identify the alkali or alkaline earth element and the halide in an unknown solution.

DISCUSSION

In 1869, the Russian chemist Dmitri Mendeleev proposed that elements in the periodic table should be arranged by increasing atomic mass. In 1913, the English physicist Harry Moseley found that the elements should actually be arranged according to increasing atomic number. The modern **periodic law** states that the properties of elements in the periodic table recur in a repeating pattern, when the elements are arranged according to *increasing atomic number*.

The elements in the periodic table are found in rows and columns. Elements in horizontal rows are called **periods**, or series. The elements in vertical columns are called **groups**, or families. Within each group of elements there are similarities in chemical properties. In this experiment, we will study three families of elements, the **alkali metals**, **alkaline earth metals**, and **halides**. We can identify the elements that belong to the same family by their similar chemical reactions.

In this experiment we will observe flame tests and solution reactions for alkali and alkaline earth elements. A **flame test** is performed by placing a small amount of solution on the coiled end of a wire. The wire is then held in a burner flame, and the flame color is observed (see Figure 7.1). For example, an orange-yellow flame shows the presence of sodium in solution.

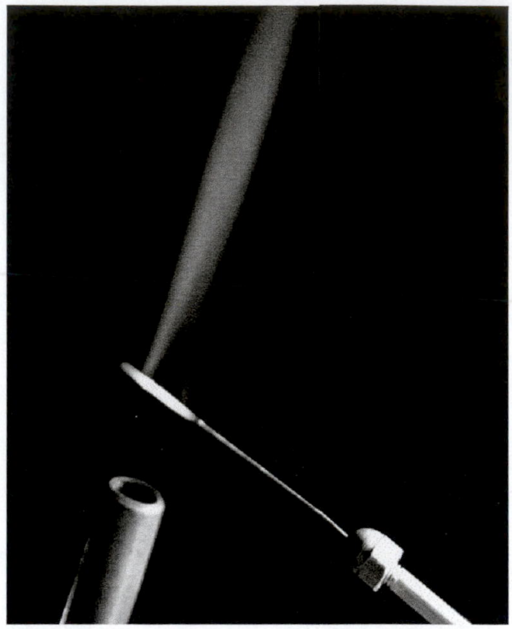

Figure 7.1 Flame-Test Technique A wire with a drop of solution in the coiled tip is placed in a hot burner flame. The color of the flame indicates the presence of a given element; a brief, yellow-green flame indicates the presence of barium; a brick-red flame indicates calcium; a scarlet-red flame indicates lithium; a violet flame indicates potassium; and a bright red flame indicates strontium.

Although the colors are specific for each element, flame tests can be misleading. Sodium is usually present as an impurity, and gives a weak yellow flame test. The intensity of the yellow flame for a sodium impurity is not as strong, and the distinction between a sodium impurity and a sodium sample can be made with a little practice.

EQUIPMENT and CHEMICALS

- 13 x 100 mm test tubes (6) & test tube rack
- test tube brush
- flame-test wire (nichrome or platinum wire)
- wash bottle with distilled water

- ammonium carbonate solution 0.5 *M* $(NH_4)_2CO_3$
- ammonium phosphate solution 0.5 *M* $(NH_4)_2HPO_4$
- ammonium sulfate solution 0.5 *M* $(NH_4)_2SO_4$

- barium solution, 0.5 *M* $BaCl_2$
- calcium solution, 0.5 *M* $CaCl_2$
- lithium solution, 0.5 *M* LiCl
- potassium solution, 0.5 *M* KCl
- sodium solution, 0.5 *M* NaCl
- strontium solution, 0.5 *M* $SrCl_2$
- bromide solution, 0.5 *M* NaBr
- chloride solution, 0.5 *M* NaCl
- iodide solution, 0.1 *M* NaI
- hexane, C_6H_{14}
- dilute nitric acid, 6 *M* HNO_3
- chlorine water (bleach)
- unknown solutions

A. Analysis of Known Solutions

1. *Flame Tests of Known Solutions*

(a) Place six test tubes in a test tube rack. Add 10 drops of barium, calcium, lithium, potassium, sodium, and strontium solutions into separate test tubes (Figure 7.2).

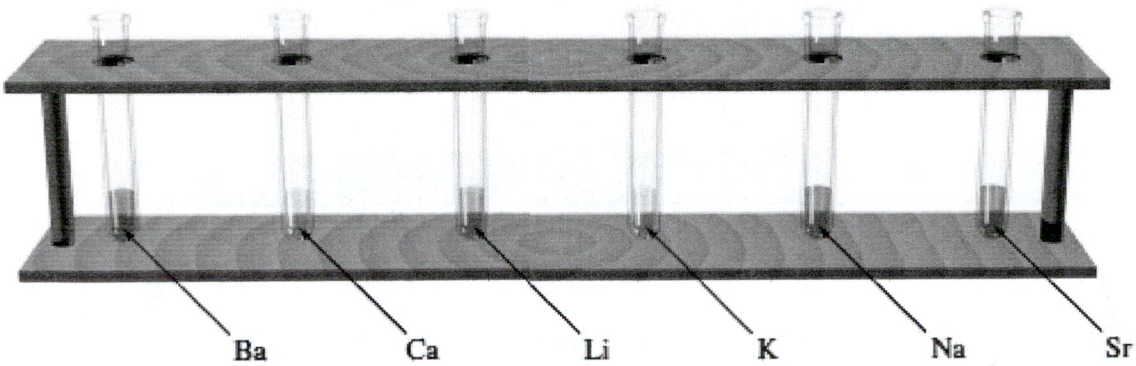

Figure 7.2 Alkali and Alkaline Earth Tests Solutions of barium, calcium, lithium, potassium, sodium, and strontium are placed in separate test tubes.

(b) Obtain a flame-test wire and make a small loop in the end. Remove contamination by placing the wire loop at the tip of a burner flame. Continue to heat the wire until there is no longer any color produced in the flame.

Note: If a flame-test wire continues to produce a colored flame, dip the wire intc dilute hydrochloric acid and heat the wire until red hot.

(c) Dip the clean flame-test wire into the test tube containing barium solution. Place the wire loop at the tip of the flame (see Figure 7.1). Record your observation. Clean the wire and repeat the flame test for calcium, lithium, potassium, sodium, and strontium solutions.

2. *Reactions of Known Solutions*

(a) Add a few drops of ammonium carbonate, $(NH_4)_2CO_3$, solution in each test tube that was used for the flame test. If a precipitate forms, record *ppt* in the Data Table. If there is no reaction, record *NR*.

(b) Clean the test tubes and rinse with distilled water. Put 10 drops of the barium, calcium, lithium, potassium, sodium, and strontium solutions into separate test tubes. Add a few drops of ammonium phosphate, $(NH_4)_3PO_4$, solution in each test tube. Record your observations in the Data Table.

(c) Clean the test tubes and put 10 drops of barium, calcium, lithium, potassium, sodium, and strontium solutions in separate test tubes. Add a few drops of ammonium sulfate, $(NH_4)_2SO_4$, in each test tube and record your observations.

3. *Halide Tests of Known Solutions*

(a) Place three test tubes in a test tube rack. Add 10 drops of bromide solution, chloride solution, and iodide solution into separate test tubes.

(b) Add 10 drops of hexane, C_6H_{14}, 1 drop of nitric acid, HNO_3, and 5 drops of chlorine water to each test tube (Figure 7.3).

(c) Shake each test tube and observe the color of the upper hexane layer.

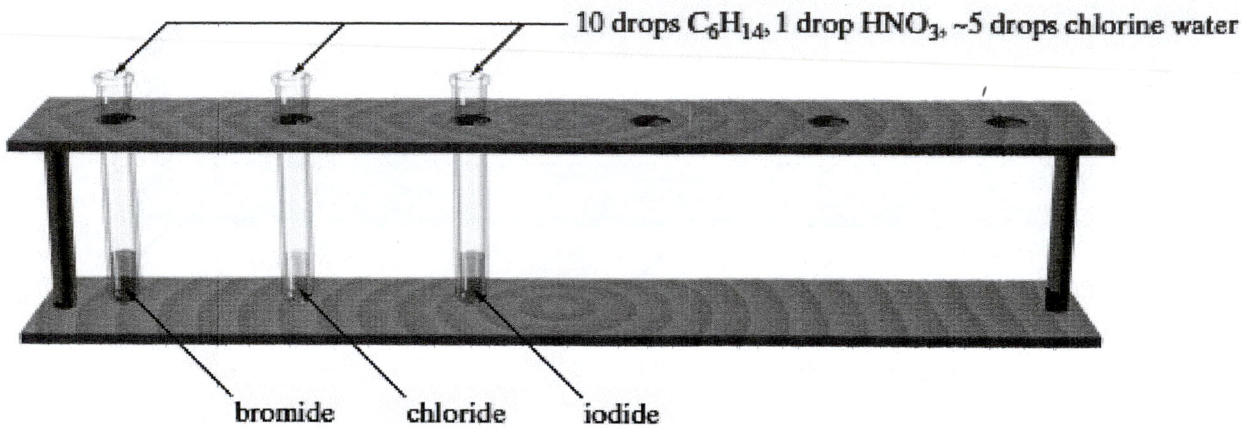

Figure 7.3 Halide Tests Hexane, nitric acid, and chlorine water are added to 10 drops of bromide, chloride, and iodide solutions in separate test tubes.

RECYCLE Chemical Waste

Note: The color of the upper hexane layer confirms the presence of bromide, chloride, or iodide. An orange color indicates bromide; a clear color indicates chloride; and a purple color indicates iodide.

Note: Dispose of the hexane layer in the organic waste container.

B. Analysis of an Unknown Solution

1. *Flame Test* Record the unknown number of a solution assigned by the Instructor. Perform a flame test on the solution, and record your observation in the Data Table.

2. *Solution Reactions* Put 10 drops of unknown solution into three test tubes. Add a few drops of ammonium carbonate, $(NH_4)_2CO_3$, to the first test tube; a few drops of ammonium phosphate, $(NH_4)_3PO_4$, to the second; and a few drops of ammonium sulfate, $(NH_4)_2SO_4$, to the third. Record your observations in the Data Table.

3. *Halide Test* Put 10 drops of unknown solution into a test tube. Add 10 drops of hexane, C_6H_{14}, 1 drop of nitric acid, HNO_3, and 5 drops of chlorine water. Shake the test tube and record the color of the upper hexane layer.

EXPERIMENT 7 NAME ____________________

DATE ____________________ SECTION ____________________

PRELABORATORY ASSIGNMENT*

1. Provide the key term that corresponds to each of the following definitions.

____________ (a) the properties of the elements recur in a repeating pattern when arranged according to increasing atomic number

____________ (b) a horizontal row of elements in the periodic table

____________ (c) a vertical column of elements in the periodic table having similar properties

____________ (d) any Group IA/1 element in the periodic table, excluding hydrogen

____________ (e) any Group IIA/2 element in the periodic table

____________ (f) a negatively charged Group VIIA/17 atom; e.g., bromide, chloride, iodide

____________ (h) a means of identifying an element by observing the characteristic color it emits when placed in a hot flame

____________ (h) an insoluble solid substance produced from a reaction in aqueous solution

____________ (i) refers to liquids that do not dissolve in one another and separate into two layers

Key Terms: alkali metal, alkaline earth metal, flame test, group, halide, immiscible, period, periodic law, precipitate (ppt)

2. Which three alkali elements are investigated in this experiment?

Which three alkaline earth elements are investigated in this experiment?

3. Refer to **Figure 7.1** to answer the following.

(a) Which element is indicated by a brief *yellow-green* flame test? ____________

(b) Which element is indicated by a *brick-red* flame test? ____________

(c) Which element is indicated by a *scarlet-red* flame test? ____________

(d) Which element is indicated by a *violet* flame test? ____________

(e) Which element is indicated by a *strong yellow* flame test? ____________

(f) Which element is indicated by a *bright red* flame test? ____________

---

* *Answers in Appendix J*

4. Where is the end of the wire placed when performing a flame test?

5. Which element occurs as an impurity and gives a weak yellow flame test?

6. Which three halides are investigated in this experiment?

7. Is water and hexane *miscible* or *immiscible*?

8. Is the halide test observed in the *upper layer* or *lower layer*?

9. Refer to **Figure 7.3** to answer the following.

   (a) Which halide is indicated by an *orange* upper layer? ______________

   (b) Which halide is indicated by a *clear* upper layer? ______________

   (c) Which halide is indicated by a *purple* upper layer? ______________

10. What safety precautions should be observed in this experiment?

EXPERIMENT 7 NAME ______________________

DATE ______________ SECTION ______________________

DATA TABLE

A. Analysis of Known Solutions

1. *Flame Tests of Known Solutions*

Solution Tested	*Flame Test Observations*
barium solution	
calcium solution	
lithium solution	
potassium solution	
sodium solution	
strontium solution	

2. *Reactions of Known Solutions*

	Solution Reaction Observations		
Solution Tested	ammonium carbonate	ammonium phosphate	ammonium sulfate
barium solution			
calcium solution			*
lithium solution			
potassium solution			
sodium solution			
strontium solution			

* Heat gently if NR.

RECYCLE Chemical Waste

3. *Halide Tests of Known Solutions*

Solution Tested	*Hexane Layer Observations*
bromide solution	
chloride solution	
iodide solution	

B. Analysis of an Unknown Solution **UNKNOWN #** _______

1. *Flame Test of an Unknown Solution*

Solution Tested	*Flame Test Observation*
unknown solution	________________________

2. *Reactions of an Unknown Solution*

	Solution Reaction Observations		
Solution Tested	ammonium carbonate	ammonium phosphate	ammonium sulfate
unknown solution	__________	__________	__________

RECYCLE Chemical Waste

Compare the flame test and solution reactions for the unknown to the observations in Procedures A.1 and A.2. Identify the alkali or the alkaline earth element present in the unknown solution; **circle** one of the following.

barium **calcium** **lithium** **potassium** **sodium** **strontium**

3. *Halide Test of an Unknown Solution*

Solution Tested	*Hexane Layer Observation*
unknown solution	________________________

Compare the hexane layer observation for the unknown solution to the observations in Procedure A.3. Identify the halide present in the unknown solution; **circle** one of the following.

bromide **chloride** **iodide**

EXPERIMENT 7 NAME ____________________

DATE ____________ SECTION ____________________

POSTLABORATORY ASSIGNMENT

1. Refer to Data Table A.2 and state the solutions that produce reactions similar to:

   (a) Li ____________ (b) Ba ____________

2. State the elements in this experiment that belong to:

   (a) Group IA/1 ____________ (b) Group IIA/2 ____________

3. An unknown solution gives a brief green flame test. The unknown gives a white precipitate with ammonium carbonate, ammonium phosphate, and ammonium sulfate. The halide test produces a purple color in the upper hexane layer. Identify (a) the alkali or alkaline earth element, and (b) the halide present in the unknown solution.

   (a) ____________________ (b) ____________________

4. An unknown solution gives a scarlet-red flame test. The unknown gives no reaction with ammonium carbonate, ammonium phosphate, and ammonium sulfate. The halide test produces an orange color in the upper hexane layer. Identify (a) the alkali or alkaline earth element, and (b) the halide present in the unknown solution.

   (a) ____________________ (b) ____________________

5. An unknown solution gives a brick-red flame test with flashes of yellow. The unknown gives a white precipitate with ammonium carbonate and ammonium phosphate, but no reaction with ammonium sulfate. The halide test produces no reaction in the upper hexane layer. Identify (a) the alkali or alkaline earth element, and (b) the halide present in the unknown solution.

   (a) ____________________ (b) ____________________

6. Fireworks are produced by gunpowder and chemicals in a rocket shell that is fired into the air and exploded. Based on your experimental observations, which elements could produce the following colors of fireworks?

   (a) bright red ____________ (b) yellow ____________

7. In groups of elements, the metallic character (*increases / decreases*) up a group.
   In groups of elements, the atomic radius (*increases / decreases*) up a group.

8. In periods of elements, the metallic character (*increases / decreases*) left to right.
   In periods of elements, the atomic radius (*increases / decreases*) left to right.

9. Refer to the periodic table on the inside front cover of this lab manual. Select the symbol of the element that corresponds to the following description.

(a) the metal in Period 3, Group IA/1 ________________

(b) the semimetal in Period 2, Group IIIA/13 ________________

(c) the nonmetal in Period 1, Group VIIIA/18 ________________

(d) the alkali metal in Period 4 ________________

(e) the alkaline earth metal in Period 4 ________________

(f) the halogen in Period 4 ________________

(g) the noble gas in Period 4 ________________

(h) the representative element in Period 4, Group IVA/14 ________________

(i) the transition element in Period 4, Group IVB/4 ________________

(j) the rare earth element in Period 4 ________________

(k) the radioactive element in Period 5 ________________

(l) the lowest atomic mass lanthanide ________________

(m) the lowest atomic number actinide ________________

(n) the first transuranium element ________________

(o) the element with atomic mass 196.97 amu ________________

(p) the element with atomic number 80 ________________

(q) the element with mass number 222 ________________

(r) the representative element in Period 4 with three valence electrons ________________

(s) the representative element in Period 4 with six valence electrons ________________

(t) the nonmetal in Group IA/1 ________________

10. (optional) Refer to the periodic table on the inside front cover of this lab manual. Which two elements in the fourth period violate the original periodic law as stated by Mendeleev?

EXPERIMENT

8

Identifying Cations in Solution

OBJECTIVES

- To observe the chemical behavior of barium, calcium, and magnesium ions.
- To analyze an unknown solution for one or more of the following cations: Ba^{2+}, Ca^{2+}, and Mg^{2+}.
- To develop the following laboratory skills: centrifuging, flame testing, and using litmus paper.

DISCUSSION

Qualitative analysis is a systematic procedure for the separation and identification of ions present in an unknown solution. Cation analysis involves the separation and identification of each positively charged **cation** present in a sample.

If we have an **aqueous solution** containing different cations, it is possible to select a reagent that will form a **precipitate** with one of the cations, but not with the others. We can then use a **centrifuge** to separate the solid particles of precipitate from the aqueous solution. Thus, we separate the cation in the precipitate from the other cations in the original aqueous solution.

For example, we can separate the cations in a solution containing Ba^{2+}, Ca^{2+}, and Mg^{2+}, using ammonium sulfate. The sulfate ion, SO_4^{2-}, precipitates Ba^{2+}, but gives no reaction with either Ca^{2+} or Mg^{2+} cations (see Figure 8.1).

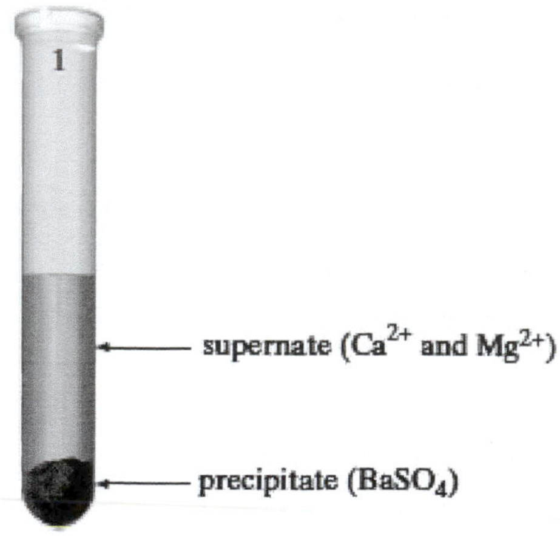

Figure 8.1 Precipitation of $BaSO_4$ The reaction of Ba^{2+} and SO_4^{2-} gives $BaSO_4$ precipitate. There is *no reaction* between Ca^{2+} and SO_4^{2-}, or Mg^{2+} and SO_4^{2-}, because $CaSO_4$ and $MgSO_4$ are both soluble.

When barium ions and sulfate ions are in a solution, a white precipitate forms because barium sulfate, $BaSO_4$, is insoluble. If calcium and sulfate ions are in a solution, no precipitate forms because calcium sulfate, $CaSO_4$, is soluble. Similarly, no precipitate forms with magnesium and sulfate ions because magnesium sulfate, $MgSO_4$, is soluble.

A **flame test** can be used to confirm the presence of an ion. A flame test is performed by dipping a wire into a solution and holding the wire in a flame while observing the color produced (Figure 8.2). Different ions produce different colored flames. For example, sodium ions give an orange-yellow flame. Since sodium is usually present as an impurity, flame tests are invariably contaminated by the orange-yellow sodium flame.

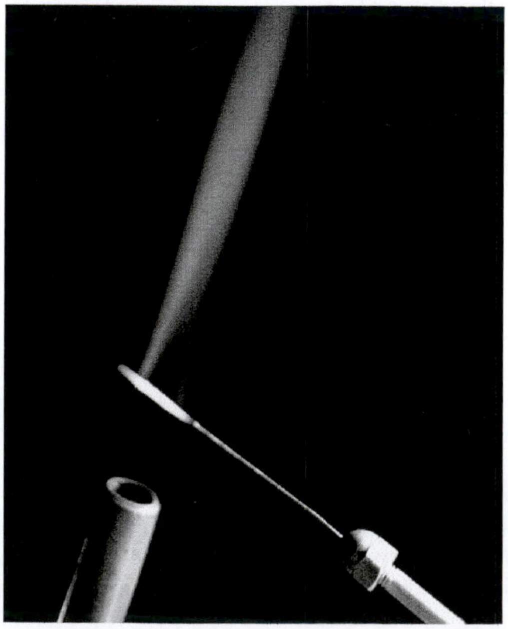

Figure 8.2 Flame-Test Technique A wire with a drop of solution in the coiled tip is placed in a hot burner flame. A brief, yellow-green flame indicates barium; and a brick-red flame indicates calcium.

Litmus paper can be used to determine whether a solution is acidic or basic. A glass stirring rod is placed in the solution and touched to the litmus paper. Acidic solutions make a *red spot* on blue litmus paper. Basic solutions make a ***blue spot*** on red litmus paper (Figure 8.3).

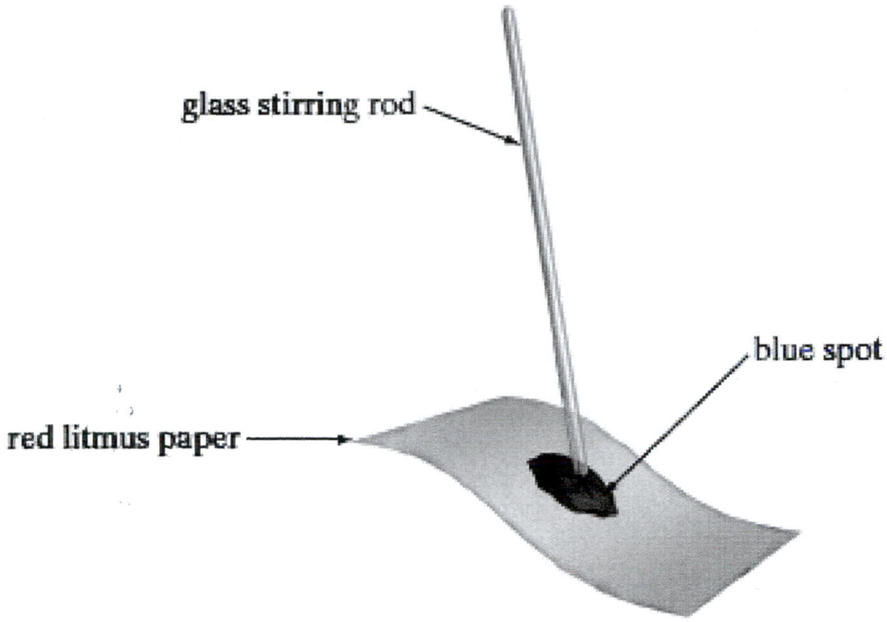

Figure 8.3 Litmus Paper Test for a Basic Solution A glass stirring rod is placed in a solution and touched to red litmus paper. A ***blue spot*** appears if the solution is basic. If there is no change, the solution is neutral or acidic.

In this experiment, you will identify Ba^{2+}, Ca^{2+}, and Mg^{2+}. First, a known solution containing all three cations will be analyzed to develop the necessary techniques. Second, an unknown solution containing one or more cations will be analyzed to determine the cations present.

EQUIPMENT and CHEMICALS

- 13 x 100 mm test tubes (3) & test tube rack
- thin glass stirring rod
- wash bottle with distilled water
- test tube brush
- red litmus paper
- flame-test wire
- known cation solution (Ba^{2+}, Ca^{2+}, and Mg^{2+} as 0.1 *M* $BaCl_2$, $CaCl_2$, $MgCl_2$)

- ammonium sulfate solution, 0.1 *M* $(NH_4)_2SO_4$
- ammonium oxalate solution, 0.1 *M* $(NH_4)_2C_2O_4$
- sodium hydrogen phosphate, 0.1 *M* Na_2HPO_4
- magnesium indicator (0.1 g *para*-nitrobenzene-azo-resorcinol in 1 L of 0.025 *M* NaOH)
- dilute hydrochloric acid, 6 *M* HCl
- dilute sodium hydroxide, 6 *M* NaOH
- unknown cation solutions (Ba^{2+}, Ca^{2+}, and/or Mg^{2+} as 0.1 *M* $BaCl_2$, $CaCl_2$, $MgCl_2$)

PROCEDURE

General Directions: Clean three test tubes and a glass stirring rod with distilled water. Label the test tubes #1, #2, and #3.

A. Analysis of a Known Cation Solution

1. *Identification of Ba^{2+} in a Known Solution*

(a) Place 10 drops of the known solution in test tube #1. Add 20 drops of ammonium sulfate, $(NH_4)_2SO_4$, and mix with a glass stirring rod.

Note: A white precipitate suggests Ba^{2+} is present.

(b) Centrifuge, and add 1 drop of ammonium sulfate to verify complete precipitation. Pour off the supernate into test tube #2, and save for Step 2.

(c) Add 5 drops of dilute hydrochloric acid, HCl, to test tube #1, and stir thoroughly. Clean a flame-test wire with hydrochloric acid and dip it into the solution. Place the wire loop in a hot flame, and record the color.

Note: A green flame test confirms Ba^{2+} is present.

2. *Identification of Ca^{2+} in a Known Solution*

RECYCLE
Chemical
Waste

(a) Add 10 drops of ammonium oxalate, $(NH_4)_2C_2O_4$, to the solution in test tube #2.

Note: A white precipitate suggests Ca^{2+} is present.

(b) Centrifuge, and add 1 drop of ammonium oxalate to verify complete precipitation. Pour off the supernate into test tube #3, and save for Step 3.

(c) Add 5 drops of dilute hydrochloric acid to test tube #2, and stir thoroughly. Clean a flame-test wire with dilute HCl and dip the wire into the solution. Place the wire loop in a hot flame, and record the color.

Note: A brick-red flame test confirms Ca^{2+} is present.

3. *Identification of Mg^{2+} in a Known Solution*

(a) Add 10 drops of sodium hydrogen phosphate, Na_2HPO_4, to the solution in test tube #3. Add 1 drop of sodium hydroxide, NaOH, and stir with a glass rod.

Note: A white precipitate suggests Mg^{2+} is present.

(b) Centrifuge, and discard the supernate.

(c) Dissolve the precipitate with dilute hydrochloric acid in test tube #3. Add 5 drops of magnesium indicator. Add sodium hydroxide, NaOH, dropwise until the solution turns red litmus paper blue. Centrifuge the precipitate.

Note: A blue gel precipitate confirms Mg^{2+} is present.

B. Analysis of an Unknown Cation Solution

1. *Identification of Ba^{2+} in an Unknown Solution*

(a) Place 10 drops of unknown solution in test tube #1. Add 20 drops of ammonium sulfate, $(NH_4)_2SO_4$, and stir with a glass rod.

Note: If there is no precipitate, Ba^{2+} is absent. Go directly to Step 2.

(b) Centrifuge, and then test for completeness of precipitation by adding 1 drop of ammonium sulfate. Pour off the supernate into test tube #2, and save for Step 2.

(c) Add 5 drops of dilute hydrochloric acid, HCl, to test tube #1 and stir thoroughly. Clean a flame-test wire with dilute HCl, and dip the wire into the solution. Place the wire loop in a hot flame, and record the color.

Note: A green flame test confirms Ba^{2+} is present.

2. *Identification of Ca^{2+} in an Unknown Solution*

(a) Add 10 drops of ammonium oxalate, $(NH_4)_2C_2O_4$, to the solution in test tube #2.

Note: If there is no precipitate, Ca^{2+} is absent. Go directly to Step 3.

RECYCLE
Chemical
Waste

(b) Centrifuge, and then test for completeness of precipitation by adding 1 drop of ammonium oxalate. Pour off the supernate into test tube #3, and save for Step 3.

(c) Add 5 drops of dilute hydrochloric acid, HCl, to test tube #2 and stir thoroughly. Clean a flame-test wire with dilute HCl, and dip the wire into the solution. Place the wire loop in a hot flame, and record the color.

Note: A brick-red flame test confirms Ca^{2+} is present.

3. *Identification of Mg^{2+} in an Unknown Solution*

(a) Add 10 drops of sodium hydrogen phosphate, Na_2HPO_4, to the solution in test tube #3. Add 1 drop of sodium hydroxide, NaOH, and stir with a glass rod.

Note: If there is no precipitate, Mg^{2+} is absent.

(b) Centrifuge, and discard the supernate.

(c) Dissolve the precipitate with dilute hydrochloric acid in test tube #3. Add 5 drops of magnesium indicator. Add sodium hydroxide, NaOH, dropwise until the solution turns red litmus paper blue. Centrifuge the precipitate.

Note: A blue gel precipitate confirms Mg^{2+} is present.

4. Based on the observations in steps 1–3, identify the cation(s) present in the unknown solution.

EXPERIMENT 8 NAME ______________________

DATE ______________ SECTION ______________________

PRELABORATORY ASSIGNMENT*

1. Provide the key term that corresponds to each of the following definitions.

___________ (a) any positively charged ion

___________ b) a systematic procedure for the separation and identification of cations, or other substances present in a sample

___________ (c) a solution of a substance dissolved in water

___________ (d) an insoluble solid substance produced from a reaction in aqueous solution

___________ (e) an instrument that spins test tubes to separate a precipitate from solution

___________ (f) the solution above a precipitate after insoluble particles are centrifuged from solution

___________ (g) the process of pouring a liquid from one container into another

___________ (h) a means of identifying an ion by observing the characteristic color it emits when placed in a hot flame

Key Terms: aqueous solution, cation, centrifuge, decant, flame test, precipitate (ppt), qualitative analysis, supernate

2. Which three cations are investigated in this experiment? (Refer to **Figure 8.1**.)

3. Where is the end of the wire placed when performing a flame test? (Refer to **Figure 8.2**.)

4. What color is the spot on red litmus paper when testing a basic solution? (Refer to **Figure 8.3**.)

5. Refer to the **Data Table** to answer the following.

   (a) Which cation is confirmed in test tube #1 by a *yellow-green* flame test? __________

   (b) Which cation is confirmed in test tube #2 by a *brick-red* flame test? __________

   (c) Which cation is confirmed in test tube #3 by a *blue "lake"* precipitate? __________

---

* *Answers in Appendix J*

6. Why is it necessary to use distilled water throughout the experiment?

7. Refer to the **Data Table** to determine which cations (Ba^{2+}, Ca^{2+}, Mg^{2+}) are present and absent in an unknown solution given the following observations.

- The unknown solution in test tube #1 plus $(NH_4)_2SO_4$ gives a white precipitate.
- The supernate in test tube #1 is poured into test tube #2.
- The white precipitate in test tube #1 gives a green flame test.
- The solution in test tube #2 plus $(NH_4)_2C_2O_4$ gives a white precipitate.
- The white precipitate in test tube #2 gives a brick-red flame test.
- The supernate in test tube #2 is poured into test tube #3.
- The solution in test tube #3 plus Na_2HPO_4 and NaOH gives a white precipitate.
- The white precipitate in test tube #3 dissolves in HCl; magnesium indicator and NaOH is added until the solution tests basic. A blue gel is observed at the bottom of the test tube.

Cation(s) **present** ____________ Cation(s) **absent** ____________

8. Refer to the **Data Table** to determine which cations (Ba^{2+}, Ca^{2+}, Mg^{2+}) are present and absent in an unknown solution given the following observations.

- The unknown solution in test tube #1 plus $(NH_4)_2SO_4$ gives no reaction.
- The solution in test tube #1 is poured into test tube #2.
- The solution in test tube #2 plus $(NH_4)_2C_2O_4$ gives no reaction.
- The solution in test tube #2 is poured into test tube #3.
- The solution in test tube #3 plus Na_2HPO_4 and NaOH gives a white precipitate.
- The white precipitate in test tube #3 dissolves in HCl; magnesium indicator and NaOH is added until the solution tests basic. A blue gel is observed at the bottom of the test tube.

Cation(s) **present** ____________ Cation(s) **absent** ____________

9. What safety precautions should be taken while performing this experiment?

EXPERIMENT 8 NAME ______________________

DATE ______________ SECTION ______________________

DATA TABLE

A. Analysis of a Known Cation Solution

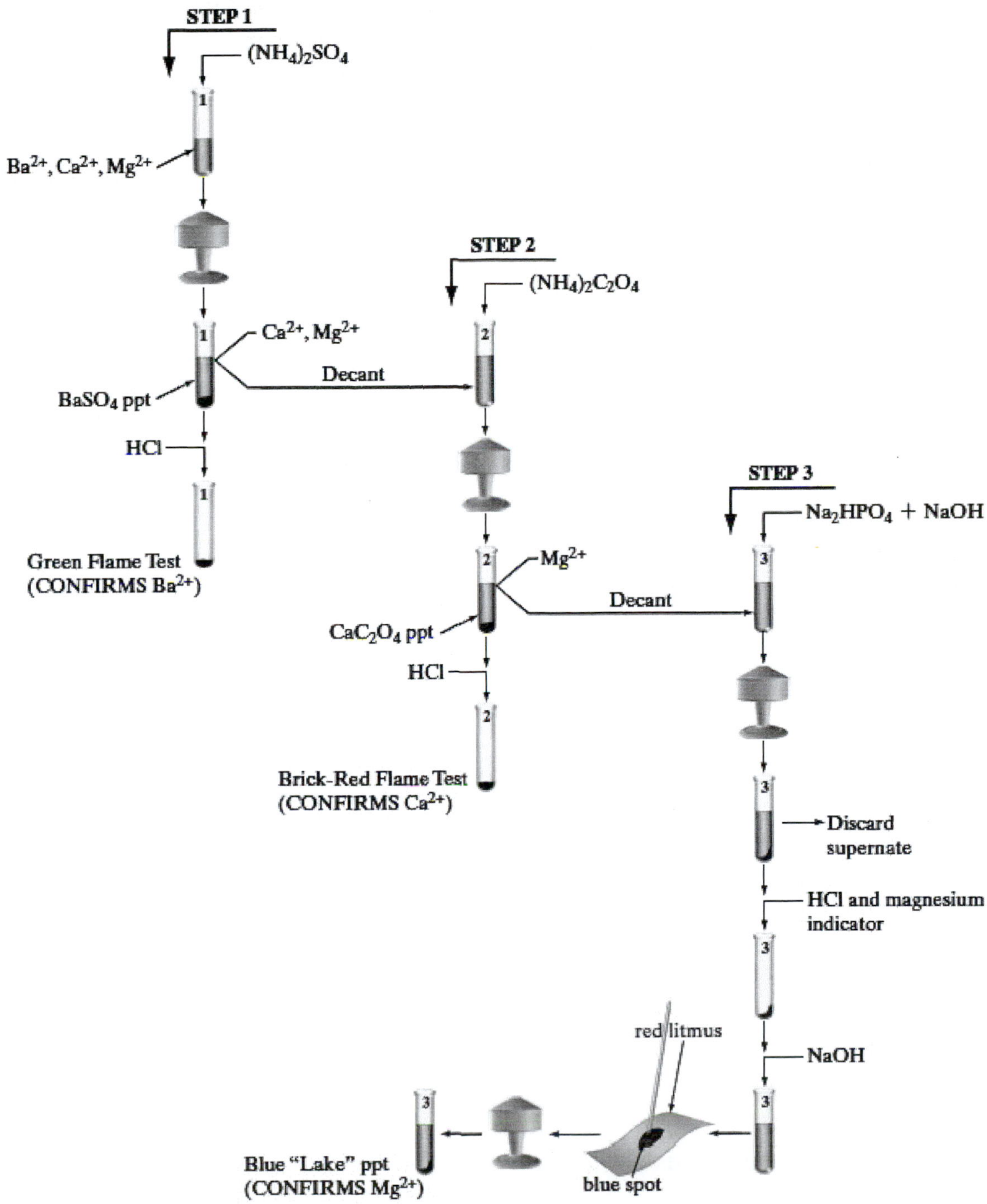

RECYCLE
Chemical
Waste

B. Analysis of an Unknown Cation Solution **UNKNOWN #** _______

Cation(s) **present** ____________ Cation(s) **absent** ____________

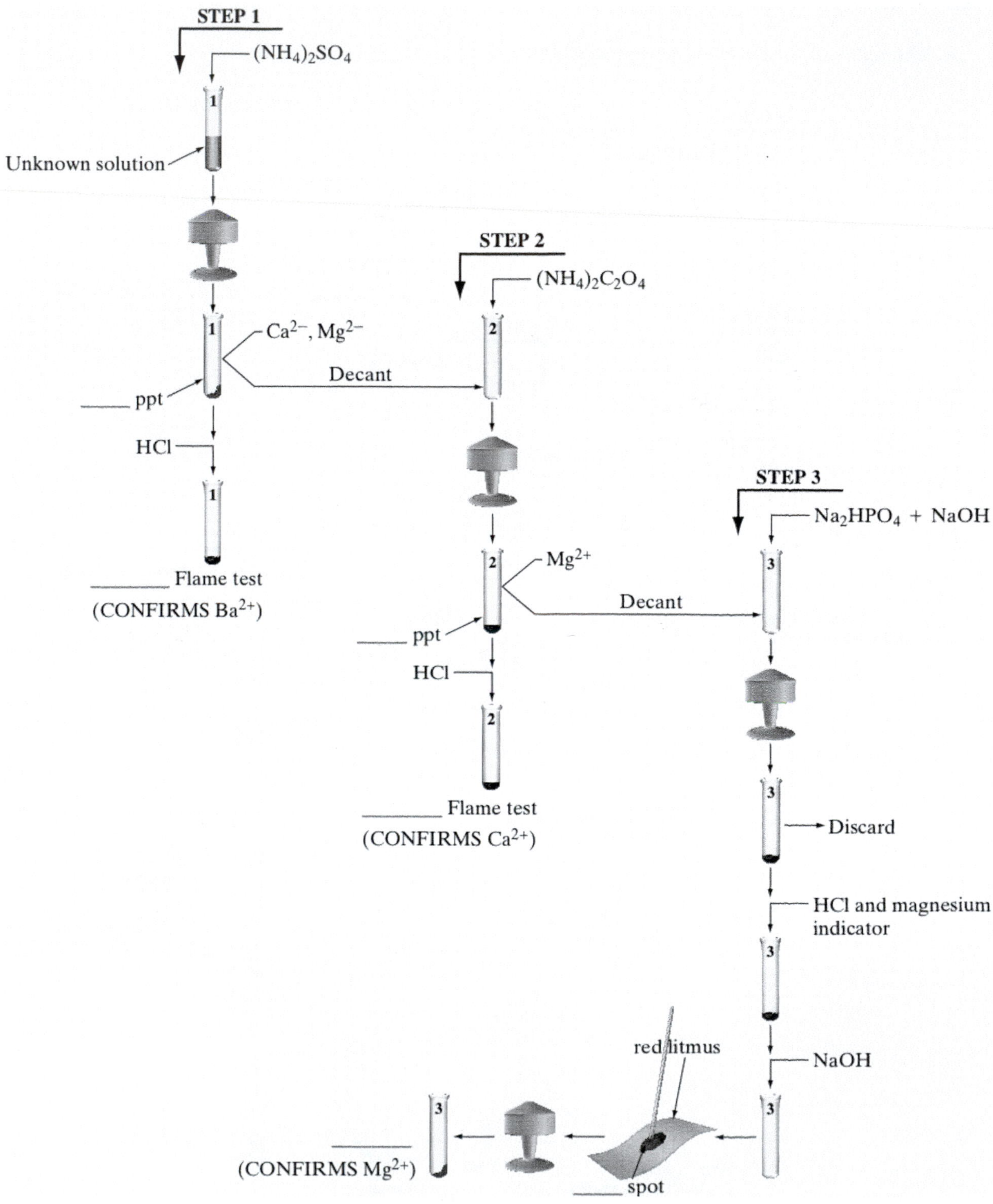

RECYCLE
Chemical
Waste

EXPERIMENT 8 NAME ______________________

DATE ______________ SECTION ______________________

POSTLABORATORY ASSIGNMENT

1. Which cations are present and absent in an unknown solution given the following observations?

   - The unknown solution in test tube #1 plus $(NH_4)_2SO_4$ gives a white precipitate.
   - The supernate in test tube #1 is poured into test tube #2.
   - The white precipitate in test tube #1 gives a green flame test.
   - The solution in test tube #2 plus $(NH_4)_2C_2O_4$ gives a white precipitate.
   - The white precipitate in test tube #2 gives a brick-red flame test.
   - The supernate in test tube #2 is poured into test tube #3.
   - The solution in test tube #3 plus Na_2HPO_4 and NaOH gives no reaction.

   Cation(s) **present** ____________ Cation(s) **absent** ____________

2. Which cations are present and absent in an unknown solution given the following observations?

   - The unknown solution in test tube #1 plus $(NH_4)_2SO_4$ gives no reaction.
   - The solution in test tube #1 is poured into test tube #2.
   - The solution in test tube #2 plus $(NH_4)_2C_2O_4$ gives a white precipitate.
   - The white precipitate in test tube #2 gives a brick-red flame test.
   - The supernate in test tube #2 is poured into test tube #3.
   - The solution in test tube #3 plus Na_2HPO_4 and NaOH gives a white precipitate.
   - The white precipitate in test tube #3 dissolves in HCl; magnesium indicator and NaOH is added until the solution tests basic. A blue gel is observed at the bottom of the test tube.

   Cation(s) **present** ____________ Cation(s) **absent** ____________

3. Write the **Stock system** name for the following cations.

   (a) Cu^+ ______________ (b) Cu^{2+} ______________

   (c) Fe^{2+} ______________ (d) Fe^{3+} ______________

   (e) Sn^{2+} ______________ (f) Sn^{4+} ______________

4. Write the **Latin system** name for the following cations.

   (a) Cu^+ ______________ (b) Cu^{2+} ______________

   (c) Fe^{2+} ______________ (d) Fe^{3+} ______________

   (e) Sn^{2+} ______________ (f) Sn^{4+} ______________

5. Complete the table below as shown by the example. Combine the ions into a correct formula, and name the compound.

	NO_3^-	SO_4^{2-}	PO_4^{3-}
Ba^{2+}	$Ba(NO_3)_2$ ***barium nitrate***		
Ca^{2+}			
Mg^{2+}			

6. (optional) Complete the table below as shown by the example. Combine the ions into a correct formula, and name the compound.

	nitrite ion	sulfite ion	phosphite ion
mercury(II) ion	$Hg(NO_2)_2$ ***mercury(II) nitrite***		
iron(III) ion			
lead(IV) ion			

Identifying Anions in Solution

EXPERIMENT

9

OBJECTIVES

- To observe the chemical behavior of iodide, chloride, and sulfate ions.
- To analyze an unknown solution for one or more of the following anions: I^-, Cl^-, and SO_4^{2-}.
- To develop the following laboratory skills: centrifuging, washing a precipitate, and using litmus paper.

DISCUSSION

Qualitative analysis is a systematic procedure for the separation and identification of ions present in an unknown solution. Anion analysis involves the separation and identification of each negatively charged **anion** present in a sample.

If we have an **aqueous solution** containing different anions, it is possible to select a reagent that will form a **precipitate** with one of the anions, but not with the others. We can then use a **centrifuge** to separate the solid particles of precipitate from the aqueous solution. Thus, we separate the anion in the precipitate from the other anions in the original aqueous solution.

For example, we can separate the anions in a solution containing I^-, Cl^-, and SO_4^{2-}, using silver nitrate. The silver ion, Ag^+, precipitates I^- and Cl^-, but gives no reaction with the SO_4^{2-} anion (see Figure 9.1).

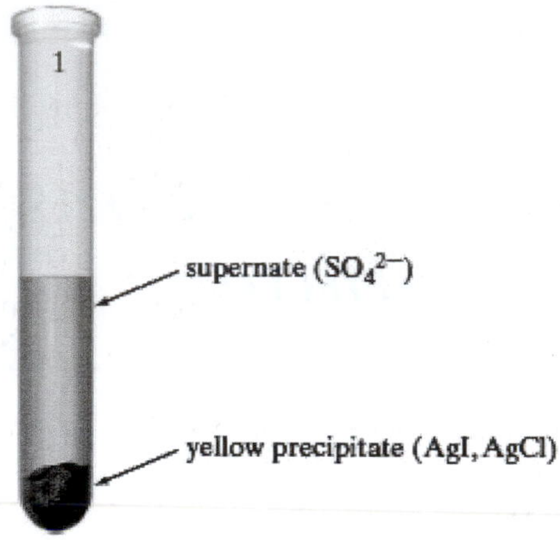

Figure 9.1 Precipitation of AgI and AgCl The reaction of Ag^+ and I^- gives AgI precipitate; Ag^+ and Cl^- gives AgCl precipitate.. There is *no reaction* between Ag^+ and SO_4^{2-} because Ag_2SO_4 is soluble.

When silver ions and iodide ions are in a solution, a yellow precipitate forms because silver iodide, AgI, is insoluble. Similarly, silver ions and chloride ions in a solution give a white precipitate of silver chloride, AgCl. If silver ions and sulfate ions are in a solution, no precipitate forms because silver sulfate, Ag_2SO_4, is soluble.

In this experiment, you will identify I^-, Cl^-, and SO_4^{2-}. First, a known solution containing all three anions will be analyzed to develop the necessary techniques. Second, an unknown solution containing one or more anions will be analyzed to determine the anions present.

Litmus paper can be used to determine whether a solution is acidic or basic. A glass stirring rod is placed in the solution and touched to the litmus paper. Basic solutions make a *blue spot* on red litmus paper. Acidic solutions make a ***red spot*** on blue litmus paper (Figure 9.2).

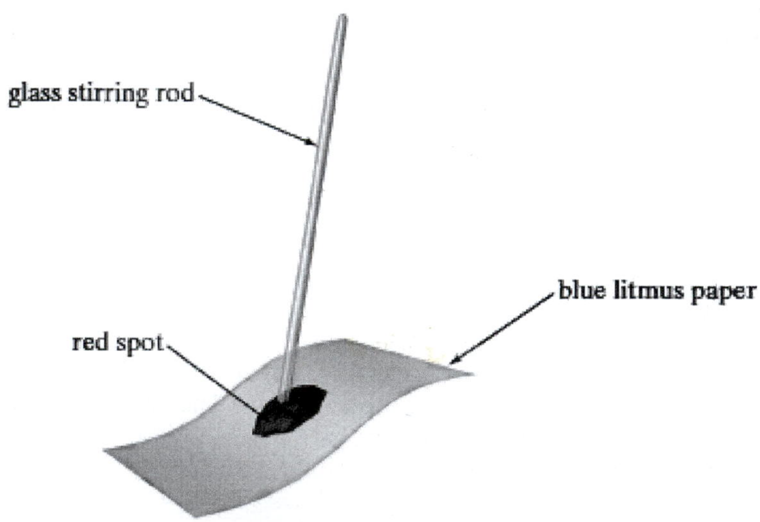

Figure 9.2 Litmus Paper Test for an Acidic Solution A glass stirring rod is placed in a solution and touched to blue litmus paper. A ***red spot*** appears if the solution is acidic. If there is no change, the solution is neutral or basic.

EQUIPMENT and CHEMICALS

- 13 x 100 mm test tubes (3) & test tube rack
- thin glass stirring rod
- wash bottle with distilled water
- centrifuge
- blue litmus paper
- silver nitrate solution, 0.1 *M* $AgNO_3$
- dilute ammonium hydroxide, 6 *M* NH_4OH
- dilute nitric acid, 6 *M* HNO_3
- barium nitrate solution, 0.1 *M* $Ba(NO_3)_2$
- known anion solution (I^-, Cl^-, and SO_4^{2-} as 0.1 *M* NaI, NaCl, Na_2SO_4)
- unknown anion solutions (I^-, Cl^-, and/or SO_4^{2-} as 0.1 *M* NaI, NaCl, Na_2SO_4)

PROCEDURE

General Directions: Clean three test tubes and a glass stirring rod with distilled water. Label the test tubes #1, #2, and #3.

A. Analysis of a Known Anion Solution

1. *Identification of I^- in a Known Solution*

(a) Place 10 drops of the known solution in test tube #1. Add 20 drops of silver nitrate, $AgNO_3$, and mix with a glass stirring rod.

Note: A yellow precipitate, AgI, suggests I^- is present.

RECYCLE Chemical Waste

(b) Centrifuge the precipitate. Pour the supernate into test tube #3 and save for Step 3.

(c) Add 10 drops of dilute ammonium hydroxide, NH_4OH, to test tube #1 and stir thoroughly with a glass rod. Centrifuge the precipitate. Pour the supernate into test tube #2 and save for Step 2.

Note: A yellow precipitate, AgI, confirms I^- is present.
(*If the precipitate is white, add 10 drops of water and stir with a glass rod.*)

2. *Identification of Cl^- in a Known Solution*

Add dilute nitric acid, HNO_3, dropwise into test tube #2 until the solution turns blue litmus paper red. Centrifuge the precipitate.

Note: A white precipitate, AgCl, confirms Cl^- is present.
(*If the precipitate is yellow, it contains AgI particles from test tube #1.*)

3. *Identification of SO_4^{2-} in a Known Solution*

Add 10 drops of barium nitrate, $Ba(NO_3)_2$, to the solution in test tube #3. Centrifuge the precipitate.

Note: A white precipitate, $BaSO_4$, confirms SO_4^{2-} is present. (*If the precipitate is yellow, it contains AgI particles from test tube #1.*)

B. Analysis of an Unknown Anion Solution

1. *Identification of I^- in an Unknown Solution*

(a) Place 10 drops of unknown solution in test tube #1. Add 20 drops of silver nitrate, $AgNO_3$, and stir with a glass rod.

Note: If there is no precipitate, I^- and Cl^- are absent. Go directly to Step 3.

RECYCLE Chemical Waste

(b) Centrifuge the precipitate. Pour the supernate into test tube #3, and save for Step 3.

(c) Add 10 drops of dilute ammonium hydroxide, NH_4OH, to test tube #1, and stir thoroughly with a glass rod. Centrifuge the precipitate. Pour the supernate into test tube #2, and save for Step 2.

Note: If there is no precipitate, I^- is absent. Go directly to Step 2.

2. *Identification of Cl^- in an Unknown Solution*

Add dilute nitric acid, HNO_3, dropwise to test tube #2 until the solution turns blue litmus paper red. Centrifuge the precipitate.

Note: If there is no precipitate, Cl^- is absent. Go directly to Step 3.

3. *Identification of SO_4^{2-} in an Unknown Solution*

Add 10 drops of barium nitrate, $Ba(NO_3)_2$, to the solution in test tube #3. Centrifuge the precipitate.

Note: If there is no precipitate, SO_4^{2-} is absent.

4. Based on the observations in steps 1–3, identify the anion(s) present in the unknown solution.

EXPERIMENT 9 NAME ______________________

DATE ______________ SECTION ______________________

*PRELABORATORY ASSIGNMENT**

1. Provide the key term that corresponds to each of the following definitions.

__________ (a) any negatively charged ion

__________ (b) a systematic procedure for the separation and identification of anions, or other substances present in a sample

__________ (c) a solution of a substance dissolved in water

__________ (d) an insoluble solid substance produced from a reaction in aqueous solution

__________ (e) an instrument that spins test tubes to separate a precipitate from solution

__________ (f) the solution above a precipitate after insoluble particles are centrifuged from solution

__________ (g) the process of pouring a liquid from one container into another

Key Terms: anion, aqueous solution, centrifuge, decant, precipitate (ppt), qualitative analysis, supernate

2. Which three anions are investigated in this experiment? (Refer to **Figure 9.1**.)

3. What color is the spot on blue litmus paper when testing an acidic solution? (Refer to **Figure 9.2**.)

4. Refer to the **Data Table** to answer the following.

   (a) Which anion is confirmed in test tube #1 by a *yellow* precipitate? __________

   (b) Which anion is confirmed in test tube #2 by a *white* precipitate? __________

   (c) Which anion is confirmed in test tube #3 by a *white* precipitate? __________

5. Why is it necessary to use distilled water throughout the experiment?

---

* *Answers in Appendix J*

6. Refer to the **Data Table** to determine which anions (I^-, Cl^-, $SO_4{}^{2-}$) are present and absent in an unknown solution given the following observations.

- The unknown solution in test tube #1 plus $AgNO_3$ gives a yellow precipitate.
- The supernate in test tube #1 is poured into test tube #3.
- The yellow precipitate in test tube #1 partially dissolves in NH_4OH.
- The supernate in test tube #1 is poured into test tube #2.
- The solution in test tube #2 is made acidic with HNO_3 and gives a white precipitate.
- The solution in test tube #3 plus $Ba(NO_3)_2$ gives a white precipitate.

Anion(s) **present** ____________ Anion(s) **absent** ____________

7. Refer to the **Data Table** to determine which anions (I^-, Cl^-, $SO_4{}^{2-}$) are present and absent in an unknown solution given the following observations.

- The unknown solution in test tube #1 plus $AgNO_3$ gives no reaction.
- The solution in test tube #1 is poured into test tube #3.
- The solution in test tube #3 plus $Ba(NO_3)_2$ gives a white precipitate.

Anion(s) **present** ____________ Anion(s) **absent** ____________

8. What safety precautions should be taken while performing this experiment?

EXPERIMENT 9 NAME ______________________

DATE ______________ SECTION ______________________

DATA TABLE

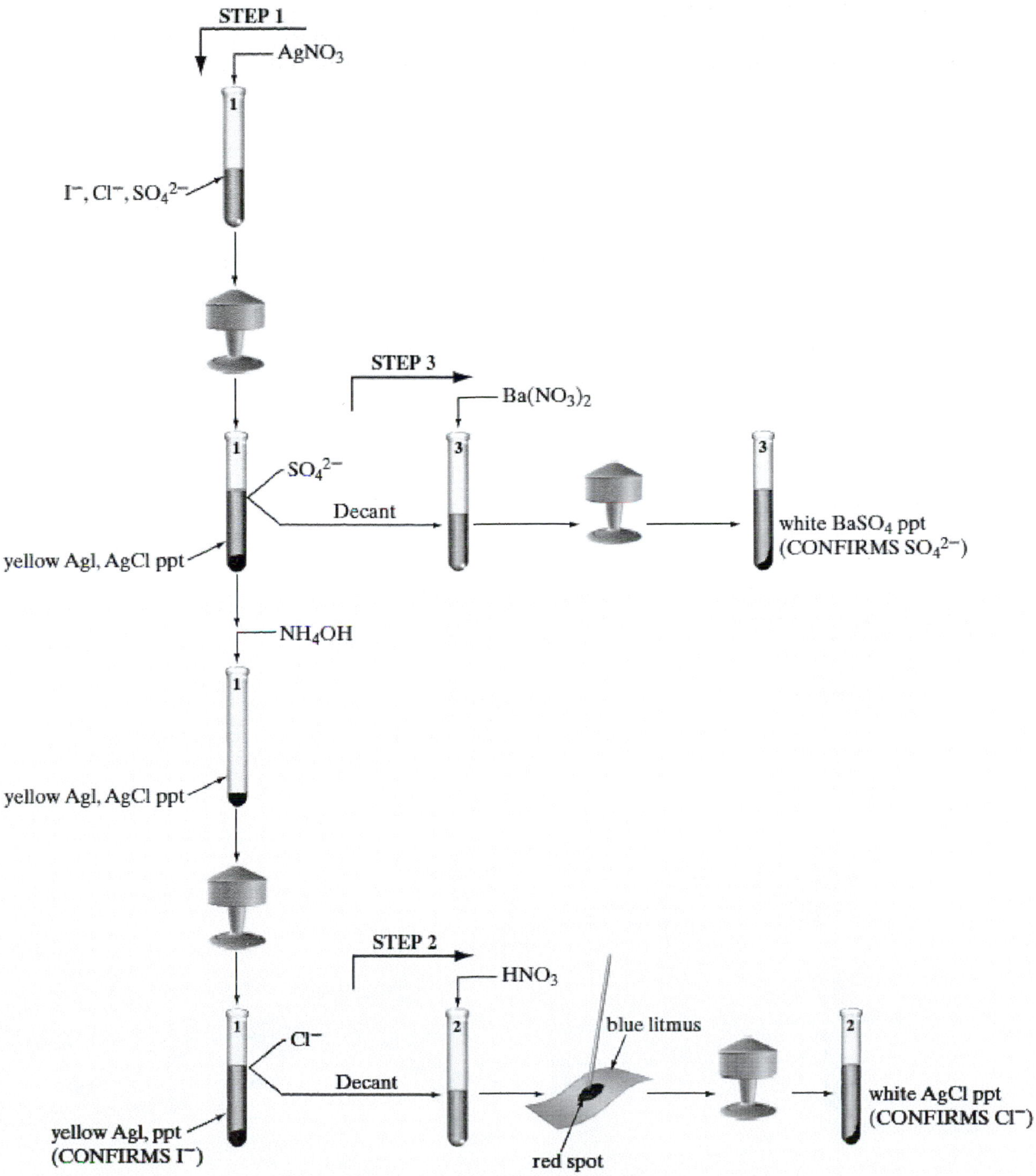

RECYCLE
Chemical
Waste

B. Analysis of an Unknown Anion Solution

UNKNOWN # _______

Anion(s) **present** ____________ Anion(s) **absent** ____________

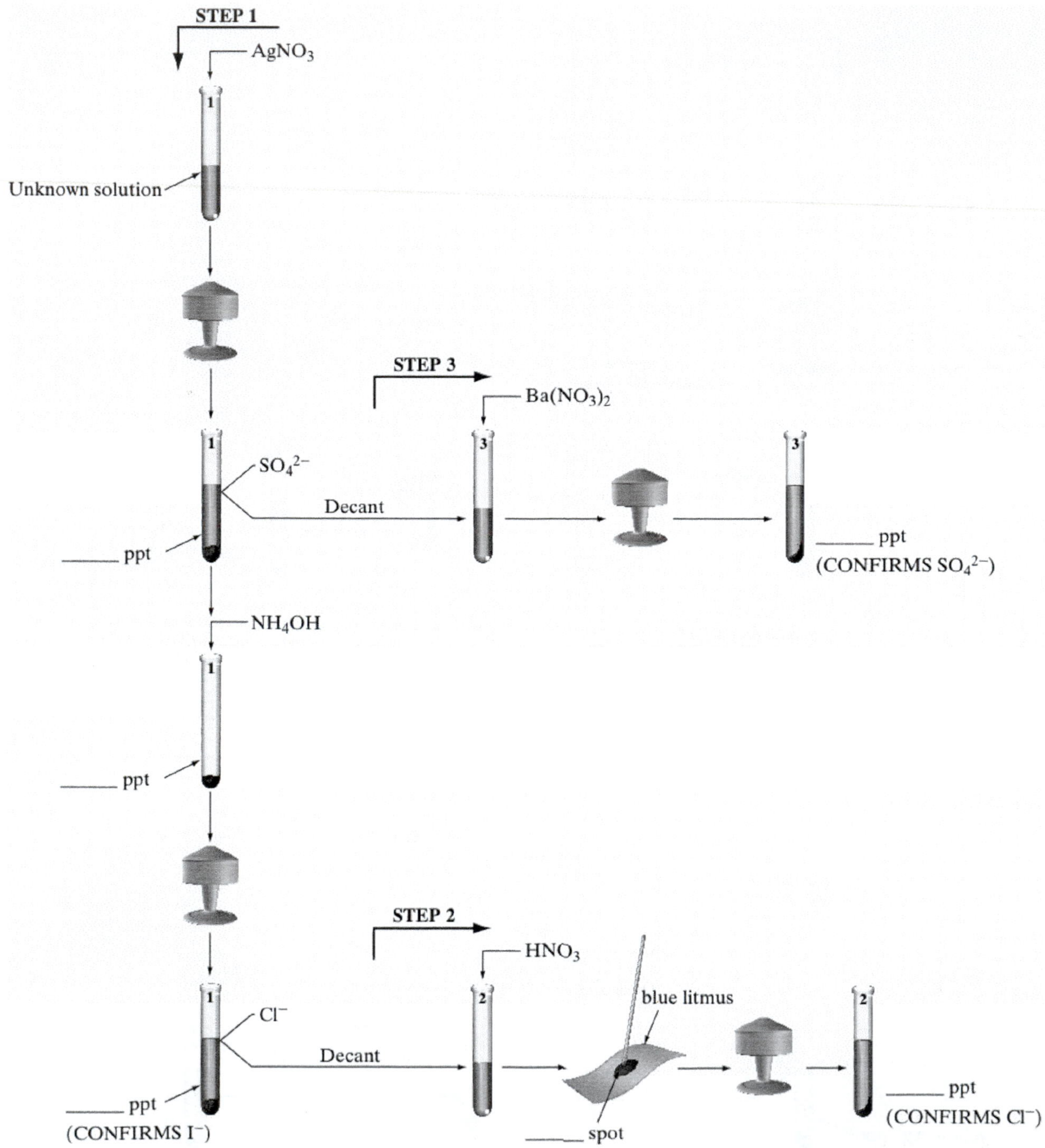

EXPERIMENT 9 NAME ______________________________

DATE ____________________ SECTION ______________________________

POSTLABORATORY ASSIGNMENT

1. Which anions are present and absent in an unknown solution given the following observations?

   - The unknown solution in test tube #1 plus $AgNO_3$ gives a yellow precipitate.
   - The supernate in test tube #1 is poured into test tube #3.
   - The yellow precipitate in test tube #1 partially dissolves in NH_4OH.
   - The supernate in test tube #1 is poured into test tube #2.
   - The solution in test tube #2 is made acidic with HNO_3 and gives a white precipitate.
   - The solution in test tube #3 plus $Ba(NO_3)_2$ gives no reaction.

   Anion(s) **present** _____________ Anion(s) **absent** _____________

2. Which anions are present and absent in an unknown solution given the following observations?

   - The unknown solution in test tube #1 plus $AgNO_3$ gives a white precipitate.
   - The supernate in test tube #1 is poured into test tube #3.
   - The white precipitate in test tube #1 dissolves completely in NH_4OH.
   - The solution in test tube #1 is poured into test tube #2.
   - The solution in test tube #2 is made acidic with HNO_3 and gives a white precipitate.
   - The solution in test tube #3 plus $Ba(NO_3)_2$ gives a white precipitate.

   Anion(s) **present** _____________ Anion(s) **absent** _____________

3. Provide the formula for the following monoatomic anions.

   (a) fluoride ion _____________ (b) bromide ion _____________

   (c) oxide ion _____________ (d) sulfide ion _____________

   (e) nitride ion _____________ (f) phosphide ion _____________

4. Provide the formula for the following polyatomic anions.

   (a) nitrate ion _____________ (b) nitrite ion _____________

   (c) sulfate ion _____________ (d) sulfite ion _____________

   (e) chlorate ion _____________ (f) chlorite ion _____________

5. Complete the table below as shown by the example. Combine the ions into a correct formula, and name the compound.

	I^-	Cl^-	SO_4^{2-}
NH_4^+	NH_4I *ammonium iodide*		
Cd^{2+}			
Al^{3+}			

6. (optional) Complete the table below as shown by the example. Combine the ions into a correct formula, and name the compound.

	acetate ion	carbonate ion	phosphate ion
copper(II) ion	$Cu(C_2H_3O_2)_2$ *copper(II) acetate*		
cobalt(III) ion			
tin(IV) ion			

EXPERIMENT

10

Analysis of a Penny

OBJECTIVES

- To state observations that are evidence for a chemical reaction.
- To write chemical equations from the descriptions of reactions.
- To determine the percentages of copper and zinc in a "zinc penny."
- To gain experience in observing chemical reactions.

DISCUSSION

Most ordinary chemical reactions can be classified as one of five basic types. The first type of reaction occurs when two or more **reactants** combine to form a single **product**. This type of reaction is called a *combination reaction.*

$$A + Z \rightarrow AZ$$

A second type of reaction occurs when a single compound breaks down into two or more simpler substances, often with the use of a **catalyst** to speed up the reaction. This type is called a *decomposition reaction.*

$$AZ \rightarrow A + Z$$

A third type of reaction occurs when one element displaces another. For this to occur, a more active element that is higher in the **activity series** displaces an element that is lower in the series. This type is called a *single-replacement reaction.*

$$A + BZ \rightarrow AZ + B$$

A fourth type of reaction occurs when two substances in **aqueous solution** switch partners; that is, an anion of one substance exchanges with another. Usually one of the products is an insoluble substance called a **precipitate**. This type is called a *double-replacement reaction.*

$$AX + BZ \rightarrow AZ + BX$$

A fifth type of reaction occurs when an acid and a base react to form a salt and water. This is a special type of double-replacement reaction, and is called a *neutralization reaction.*

$$HX + BOH \rightarrow BX + HOH$$

Notice that the hydrogen ion in the acid neutralizes the hydroxide ion in the base to form water. If water is written as HOH, the neutralization is obvious and the equation may be easier to balance.

In this experiment, you will observe evidence for a chemical reaction. Evidence for a reaction includes: (1) a *gas is released*; (2) a *precipitate is produced*; (3) a *permanent color change* is observed; (4) an *energy change* is noted, such as heat or light being given off.

In order to describe a chemical reaction, chemists use shorthand symbols when writing a chemical equation. Table 10.1 lists some of these symbols.

Table 10.1 Chemical Equation Symbols

Symbol	*Explanation of Symbol*
$\rightarrow$	produces, yields *(separates the reactants from the products)*
+	reacts with, added to *(separates two or more reactants or products)*
$\xrightarrow{\Delta}$	heat is a catalyst for the reaction
$\xrightarrow{Fe}$	iron is a catalyst for the reaction
NR	no reaction
(g)	gaseous substance
(l)	liquid substance
(s)	solid substance or precipitate
(aq)	aqueous solution

To write a chemical equation it is necessary to predict the products from a reaction. To aid you in writing equations, the products are given for each reaction in this experiment. You only need to convert the given reactions into chemical equations and balance the reactants and products. The following examples illustrate.

Combination Reaction

calcium(s) + oxygen(g) → calcium oxide(s)

$$2\ Ca(s) + O_2(g) \rightarrow 2\ CaO(s)$$

Decomposition Reaction

lithium hydrogen carbonate(s) $\xrightarrow{\Delta}$ lithium carbonate(s) + water(g) + carbon dioxide(g)

$$2\ LiHCO_3(s) \xrightarrow{\Delta} Li_2CO_3(s) + H_2O(g) + CO_2(g)$$

Single-Replacement Reaction

zinc(s) + hydrochloric acid(aq) → zinc chloride(aq) + hydrogen(g)

$$Zn(s) + 2\ HCl(aq) \rightarrow ZnCl_2(aq) + H_2(g)$$

Double-Replacement Reaction

potassium carbonate(aq) + calcium chloride(aq) → calcium carbonate(s) + potassium chloride(aq)

$$K_2CO_3(aq) + CaCl_2(aq) \rightarrow CaCO_3(s) + 2\ KCl(aq)$$

Neutralization Reaction

nitric acid(aq) + barium hydroxide(aq) → barium nitrate(aq) + water

$$2\ HNO_3(aq) + Ba(OH)_2(aq) \rightarrow Ba(NO_3)_2(aq) + 2\ HOH(l)$$

Analysis of a "Zinc Penny"

In 1982 the *U.S. Mint* stopped making copper pennies because they cost more than 1¢ to mint. The *U.S. Mint* started making pennies from zinc "blanks" plated with a thin layer of copper metal. Although copper and zinc pennies look similar, a "zinc penny" weighs about 20% less.

In this experiment you will cut a "zinc penny" to expose the zinc metal and drop the penny into sulfuric acid. Copper does not react with acid, but zinc does react and leaves a thin copper shell. The chemical equations for the two metals and sulfuric acid are:

$$Cu(s) + H_2SO_4(aq) \rightarrow NR$$

$$Zn(s) + H_2SO_4(aq) \rightarrow ZnSO_4(aq) + H_2(g)$$

The following example exercise illustrates the calculation for the percentages of copper and zinc in a "zinc penny."

Example Exercise 10.1 • Percent Composition of a "Zinc Penny"

A 1995 penny having a mass of 2.536 g is cut as shown in Figure 10.1 and dropped into sulfuric acid. After the zinc has reacted, the copper shell is found to have a mass of 0.063 g. Calculate the percentages of copper and zinc in the "zinc penny."

Solution: The percentage of copper is simply the ratio of the mass of Cu metal to the mass of the penny; that is,

$$\frac{0.063\ \cancel{g}}{2.536\ \cancel{g}} \times 100\% \quad = \quad 2.5\%\ \text{Cu}$$

The percentage of zinc is the ratio of the mass of Zn metal to the mass of the original penny. The mass of Zn corresponds to the mass loss of the penny: 2.536 g – 0.063 g = 2.473 g. Thus,

$$\frac{2.473\ \cancel{g}}{2.536\ \cancel{g}} \times 100\% \quad = \quad 97.52\%\ \text{Zn}$$

Experimentally, the 1995 "zinc penny" is 2.5% Cu and 97.52% zinc.

Students often ask if it is illegal to destroy a penny. According to a U.S. Treasury official: *"the law provides criminal penalties for anyone who **fraudulently** alters, defaces, mutilates, impairs, diminishes, falsifies, scales or lightens any of the coins coined at the mints of the United States."* Since we are not intending to defraud, this experiment is ***legal***.

EQUIPMENT and CHEMICALS

A. Instructor Demonstrations

- crucible tongs
- deflagrating spoon
- magnesium, Mg ribbon
- sulfur, S powder

B–F. Student Experiments

- ring stand & ring
- 13 x 100 mm test tubes (6) & test tube rack
- test tube holder
- test tube brush
- wash bottle with distilled water
- 250-mL Erlenmeyer flask
- 100-mL beaker
- copper(II) sulfate pentahydrate, solid $CuSO_4 \cdot 5H_2O$
- sodium hydrogen carbonate, solid $NaHCO_3$
- wooden splints
- copper, Cu wire
- magnesium, Mg turnings
- calcium, Ca turnings
- hydrochloric acid, 0.1 *M* HCl
- silver nitrate, 0.1 *M* $AgNO_3$
- copper(II) nitrate, 0.1 *M* $Cu(NO_3)_2$
- aluminum nitrate, 0.1 *M* $Al(NO_3)_3$
- potassium carbonate, 0.5 *M* K_2CO_3
- sodium phosphate, 0.5 *M* Na_3PO_4
- nitric acid, 0.1 *M* HNO_3
- sulfuric acid, 0.1 *M* H_2SO_4
- phosphoric acid, 0.1 *M* H_3PO_4
- sodium hydroxide, 0.5 *M* NaOH
- phenolphthalein indicator
- "zinc penny" (post-1982 mint date)
- dilute sulfuric acid, 3 *M* H_2SO_4
- acetone, C_3H_6O

PROCEDURE

General Directions: For Procedures A–E, record your observations in the Data Table. Since a "zinc penny" requires hours to react completely, it is advisable to start with Procedure F, and then continue with Procedures A–E.

A. Instructor Demonstration – Combination Reactions

1. Hold a 2-cm strip of magnesium ribbon with crucible tongs, and ignite the metal in a hot burner flame.

2. Put about 1 g of sulfur in a deflagrating spoon. Dim the lights and ignite the powder with a hot burner flame. Place the burning sulfur under a fume hood to avoid the strong odor of sulfur dioxide gas.

RECYCLE
Chemical
Waste

B. Decomposition Reactions

1. Put a pea-sized portion of copper(II) sulfate pentahydrate crystals into a dry test tube. Grasp the test tube with a test tube holder and heat with a burner (see Figure 1.1). Note the color change, and observe the inside surface at the top of the test tube.

2. Add sodium hydrogen carbonate (baking soda) into a 250-mL Erlenmeyer flask so as to cover the bottom of the flask. Support the flask on a ring stand, using a wire gauze.
   (a) Hold a flaming splint in the mouth of the flask, and record how long it burns.
   (b) Heat the flask strongly with the laboratory burner until moisture is observed; hold a flaming splint in the mouth of the flask, and record how long it burns.

C. Single-Replacement Reactions

1. Put 20 drops of silver nitrate solution into a test tube, and add a small piece of copper wire. Allow a few minutes for reaction and then record your observation.

2. Put 20 drops of hydrochloric acid into a test tube, and add a small piece of magnesium metal. Record your observation.

3. Put 20 drops of distilled water into a test tube, and add a small piece of calcium metal. Record your observation.

D. Double-Replacement Reactions

1–3. Put 10 drops of silver nitrate, copper(II) nitrate, and aluminum nitrate solutions into separate test tubes #1–3. Add a few drops of potassium carbonate solution to test tubes #1, #2, and #3. Observe the reactions, and record your observations.

4–6. Put 10 drops of silver nitrate, copper(II) nitrate, and aluminum nitrate solutions into separate test tubes #4–6. Add a few drops of sodium phosphate solution to test tubes #4, #5, and #6. Observe the reactions, and record your observations.

E. Neutralization Reactions

1. Put 10 drops of nitric acid, sulfuric acid, and phosphoric acid into test tubes #1–3. Add one drop of indicator to each of the test tubes. Add sodium hydroxide dropwise into test tube #1 until a permanent color change is observed.

   Note: The indicator is colorless in acid, and pink in a basic solution.

RECYCLE
Chemical
Waste

2. Add drops of dilute sodium hydroxide solution to test tube #2 until a permanent color change is observed.

3. Add drops of dilute sodium hydroxide solution to test tube #3 until a permanent color change is observed.

F. Percentages of Copper and Zinc in a Penny

1. Obtain a post-1982 penny, and record the mint date. Using metal shears, cut the coin as shown in Figure 10.1.

Figure 10.1 Exposing Zinc in a Penny A "zinc penny" should be cut as shown to ensure a rapid and complete reaction with sulfuric acid.

2. Weigh the penny on a balance, and record the mass.

3. Drop the penny into a 100-mL beaker, and add about 20 mL of dilute sulfuric acid. The reaction requires about 3 hours for the zinc in the coin to react completely.

4. When the coin stops producing gas bubbles, discard the acid, and wash the coin with distilled water.

5. Rinse the coin with acetone, and discard the rinse solution. When the coin appears dry, weigh the copper shell and record the mass in the Data Table.

6. Calculate the percentages of copper and zinc in the penny.

EXPERIMENT 10 NAME ____________________

DATE ____________ SECTION ____________________

*PRELABORATORY ASSIGNMENT**

1. Provide the key term that corresponds to each of the following definitions.

____________ (a) a substance undergoing a chemical reaction

____________ (b) a substance resulting from a chemical reaction

____________ (c) a substance that speeds up a chemical reaction

____________ (d) a relative order of metals arranged according to their ability to undergo reaction

____________ (e) a solution of a substance dissolved in water

____________ (f) an insoluble solid substance produced from a reaction in aqueous solution

____________ (g) the solution above a precipitate after insoluble particles separate from solution

Key Terms: aqueous solution, activity series, catalyst, precipitate (ppt), product, reactant, supernate

2. Classify the following types of chemical reactions.

(a) $2\ Na(s) + Cl_2(g) \rightarrow 2\ NaCl(s)$

(b) $Na_2CO_3(s) \xrightarrow{\Delta} Na_2O(s) + CO_2(g)$

(c) $2\ Na(s) + 2\ H_2O(aq) \rightarrow 2\ NaOH(aq) + H_2\ (s)$

(d) $AgNO_3(aq) + NaCl(aq) \rightarrow AgCl(s) + NaNO_3(aq)$

(e) $HNO_3(aq) + KOH(aq) \rightarrow KNO_3(aq) + H_2O(l)$

3. State four observations that provide evidence for a chemical reaction.

(a)

(b)

(c)

(d)

** Answers in Appendix J*

4. Supply the symbol for each of the following in a chemical equation.

   (a) gas reactant or product

   (b) liquid reactant or product

   (c) solid reactant or product

   (d) aqueous solution

   (e) precipitate in solution

   (f) heat as a catalyst

   (g) no reaction

5. What color is phenolphthalein indicator:

   (a) in an acidic solution?

   (b) in a basic solution?

6. Based on the mint date, which of following pennies ***cannot*** be used in this experiment: 1980, 1990, or 2000?

7. A 2010 penny has a mass of 2.541 g and produces a copper shell with a mass of 0.064 g. Refer to Example Exercise 10.1 and find the (a) %Cu and (b) %Zn in the penny.

8. What safety precautions must be observed in this experiment?

EXPERIMENT 10 NAME ______________________

DATE ______________ SECTION ______________________

DATA TABLE

Procedure	*Observation*
A. Instructor Demonstration – Combination Reactions	
1. $Mg + O_2 \xrightarrow{\Delta}$	
2. $S + O_2 \xrightarrow{\Delta}$	
B. Decomposition Reactions	
1. $CuSO_4 \cdot 5H_2O \xrightarrow{\Delta}$	
2. (a) $NaHCO_3$	
(b) $NaHCO_3 \xrightarrow{\Delta}$	
C. Single-Replacement Reactions	
1. $Cu + AgNO_3 \rightarrow$	
2. $Mg + HCl \rightarrow$	
3. $Ca + H_2O \rightarrow$	
D. Double-Replacement Reactions	
1. $AgNO_3 + K_2CO_3 \rightarrow$	
2. $Cu(NO_3)_2 + K_2CO_3 \rightarrow$	
3. $Al(NO_3)_3 + K_2CO_3 \rightarrow$	
4. $AgNO_3 + Na_3PO_4 \rightarrow$	
5. $Cu(NO_3)_2 + Na_3PO_4 \rightarrow$	
6. $Al(NO_3)_3 + Na_3PO_4 \rightarrow$	
E. Neutralization Reactions	
1. $HNO_3 + NaOH \rightarrow$	
2. $H_2SO_4 + NaOH \rightarrow$	
3. $H_3PO_4 + NaOH \rightarrow$	

RECYCLE
Chemical
Waste

Converting Word Equations into Balanced Chemical Equations

A. Instructor Demonstration – Combination Reactions

1. magnesium(s) + oxygen(g) $\xrightarrow{\Delta}$ magnesium oxide(s)

   $Mg(s) + O_2(g) \xrightarrow{\Delta}$

2. sulfur(s) + oxygen(g) $\xrightarrow{\Delta}$ sulfur dioxide(g)

   $S(s) + O_2(g) \xrightarrow{\Delta}$

B. Decomposition Reactions

1. copper(II) sulfate pentahydrate(s) $\xrightarrow{\Delta}$ copper(II) sulfate(s) + water(g)

   $CuSO_4 \cdot 5H_2O(s) \xrightarrow{\Delta}$

2. sodium hydrogen carbonate(s) $\xrightarrow{\Delta}$ sodium carbonate(s) + water(g) + carbon dioxide(g)

   $NaHCO_3(s) \xrightarrow{\Delta}$

C. Single-Replacement Reactions

1. copper(s) + silver nitrate(aq) → copper(II) nitrate(aq) + silver(s)

   $Cu(s) + AgNO_3(aq) \rightarrow$

2. magnesium(s) + hydrochloric acid(aq) → magnesium chloride(aq) + hydrogen(g)

   $Mg(s) + HCl(aq) \rightarrow$

3. calcium(s) + water(l) → calcium hydroxide(s) + hydrogen(g)

   $Ca(s) + H_2O(l) \rightarrow$

D. Double-Replacement Reactions

1. silver nitrate(aq) + potassium carbonate(aq) → silver carbonate(s) + potassium nitrate(aq)

   $AgNO_3(aq)$ + $K_2CO_3(aq)$ →

2. copper(II) nitrate(aq) + potassium carbonate(aq) → copper(II) carbonate(s) + potassium nitrate(aq)

   $Cu(NO_3)_2(aq)$ + $K_2CO_3(aq)$ →

3. aluminum nitrate(aq)+potassium carbonate(aq) → aluminum carbonate(s)+potassium nitrate(aq)

   $Al(NO_3)_3(aq)$ + $K_2CO_3(aq)$ →

4. silver nitrate(aq) + sodium phosphate(aq) → silver phosphate(s) + sodium nitrate(aq)

   $AgNO_3(aq)$ + $Na_3PO_4(aq)$ →

5. copper(II) nitrate(aq) + sodium phosphate(aq) → copper(II) phosphate(s) + sodium nitrate(aq)

   $Cu(NO_3)_2(aq)$ + $Na_3PO_4(aq)$ →

6. aluminum nitrate(aq) + sodium phosphate(aq) → aluminum phosphate(s) + sodium nitrate(aq)

   $Al(NO_3)_3(aq)$ + $Na_3PO_4(aq)$ →

E. Neutralization Reactions

1. nitric acid(aq) + sodium hydroxide(aq) → sodium nitrate(aq) + water(l)

   $HNO_3(aq)$ + $NaOH(aq)$ →

2. sulfuric acid(aq) + sodium hydroxide(aq) → sodium sulfate(aq) + water(l)

   $H_2SO_4(aq)$ + $NaOH(aq)$ →

3. phosphoric acid(aq) + sodium hydroxide(aq) → sodium phosphate(aq) + water(l)

   $H_3PO_4(aq)$ + $NaOH(aq)$ →

F. Percentages of Copper and Zinc in a Penny **Mint Date** ______

mass of "zinc penny" ____________ g

mass of copper ____________ g

mass of zinc ____________ g

Show the calculation for the percentage copper in the penny (see Example Exercise 10.1).

RECYCLE
Chemical
Waste

Percentage copper ____________ % Cu

Show the calculation for the percentage zinc in the penny.

Percentage zinc ____________ % Zn

EXPERIMENT 10 NAME ____________________

DATE ______________ SECTION ____________________

POSTLABORATORY ASSIGNMENT

1. Provide the chemical formula for the substance described in each of the chemical reactions. Refer to pages 120–121 for the substance described in each reaction.

   (a) the white smoke produced from reaction **A.1** ____________

   (b) the strong odor produced from reaction **A.2** ____________

   (c) the colorless gas produced from reaction **B.1** ____________

   (d) the flame-extinguishing gas from reaction **B.2** ____________

   (e) the gray solid produced from reaction **C.1** ____________

   (f) the colorless gas produced from reaction **C.2** ____________

   (g) the white ppt produced from reaction **C.3** ____________

   (h) the cream ppt produced from reaction **D.1** ____________

   (i) the blue-white ppt produced from reaction **D.2** ____________

   (j) the white ppt produced from reaction **D.3** ____________

   (k) the yellow ppt produced from reaction **D.4** ____________

   (l) the blue-white ppt produced from reaction **D.5** ____________

   (m) the white ppt produced from reaction **D.6** ____________

   (n) the acid reacting in reaction **E.1** ____________

   (o) the acid reacting in reaction **E.2** ____________

   (p) the acid reacting in reaction **E.3** ____________

   (q) the base reacting in reactions **E.1– E.3** ____________

2. Refer to the *Activity Series of Metals* in **Appendix F** and indicate reaction (*Rxn*) or no reaction (*NR*) when a small piece of tin metal is dropped into the following aqueous solutions.

   (a) $Ca(NO_3)_2(aq)$ ____________ (b) $Cr(NO_3)_3(aq)$ ____________

   (c) $Ni(NO_3)_2(aq)$ ____________ (d) $Pb(NO_3)_2(aq)$ ____________

   (e) $HNO_3(aq)$ ____________ (f) $AgNO_3(aq)$ ____________

3. Refer to the *Solubility Rules* in **Appendix G** and indicate whether the following compounds are soluble (*sol*) or insoluble (*insol*) in water.

(a) NH_4Cl	____________	(b) $Fe(C_2H_3O_2)_3$	____________
(c) $Mg(NO_3)_2$	____________	(d) AgI	____________
(e) $BaSO_4$	____________	(f) $CaCO_3$	____________
(g) $CuCrO_4$	____________	(h) $AlPO_4$	____________
(i) ZnS	____________	(j) $Sr(OH)_2$	____________

4. Write a balanced chemical equation for each of the following reactions.

(a) potassium(s) + chlorine(g) $\xrightarrow{\Delta}$ potassium chloride(s)

(b) calcium carbonate(s) $\xrightarrow{\Delta}$ calcium oxide(s) + carbon dioxide(g)

(c) lithium metal(s) + water(l) → lithium hydroxide(aq) + hydrogen(g)

(d) lead(II) nitrate(aq) + sodium iodide (aq) → lead(II) iodide(s) + sodium nitrate(aq)

(e) acetic acid(aq) + barium hydroxide(aq) → barium acetate(aq) + water(l)

5. (optional) A 1980 penny weighing 3.079 g reacts with nitric acid to give a blue solution. An electric current is passed through the solution and 2.925 g of copper metal is produced. What is the percentage of copper in the 1980 penny?

____

EXPERIMENT

11

Determination of Avogadro's Number

OBJECTIVES

- To determine Avogadro's number using a molecular monolayer technique.
- To find the number of molecules in a monolayer.
- To find the moles of stearic acid in a monolayer.
- To develop a sensitive technique in preparing a thin film of molecules.

DISCUSSION

Avogadro's number (symbol N) is defined as the number of carbon atoms in 12.01 g of carbon. It is an extremely large number. Avogadro's number has been determined by several experimental methods. Currently, the most precise value is 6.0221415×10^{23}.

A **mole** (symbol **mol**) is the amount of substance that contains Avogadro's number of atoms, molecules, or formula units. If we use a value of 6.02×10^{23} for Avogadro's number, we can write

$$\textit{Avogadro's number } (N) \quad = \quad 1 \textit{ mole} \quad = \quad 6.02 \times 10^{23} \textit{ particles}$$

The mass of one mole of any substance is termed its **molar mass** (symbol **MM**). The molar mass of carbon is 12.01 g, the molar mass of oxygen, O_2, is 32.00 g, and the molar mass of sodium chloride, NaCl, is 58.44 g. Since we know the mass of one mole of any substance, we also know the number of atoms corresponding to its molar mass. Thus,

$$\textit{molar mass (g/mol)} \quad = \quad 1 \textit{ mole} \quad = \quad 6.02 \times 10^{23} \textit{ particles}$$

In this experiment, we will determine an experimental value for Avogadro's number. First, a solution of stearic acid is prepared by dissolving stearic acid in hexane. The organic solution is then added one drop at a time onto the water in the watchglass.

After each drop is added, stearic acid molecules spread across the surface of the water forming a single layer; that is, a **monolayer**. A few seconds after each drop of solution is added, the solvent evaporates and the drop disappears. When enough drops of solution have been added to form a monolayer of stearic acid molecules, *one* additional drop forms a clear lens (Figure 11.1).

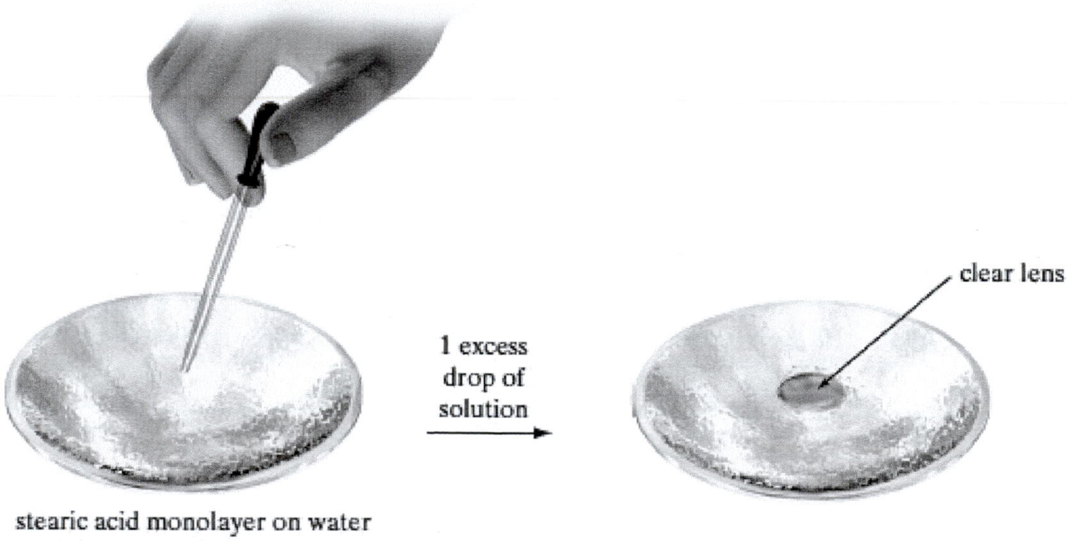

Figure 11.1 Formation of a Molecular Monolayer A single monolayer of stearic acid molecules spreads across the surface of the water in a watchglass.

The stearic acid molecule, $C_{17}H_{35}COOH$, is a long-chain molecule having a polar "head" and a nonpolar "tail." The structure of the molecule is as follows:

$$CH_3\text{-}CH_2\text{-}CH_2\text{-}CH_2\text{-}CH_2\text{-}CH_2\text{-}CH_2\text{-}CH_2\text{-}CH_2\text{-}CH_2\text{-}CH_2\text{-}CH_2\text{-}CH_2\text{-}CH_2\text{-}CH_2\text{-}CH_2\text{-}CH_2\text{-}COOH$$

Nonpolar "tail" Polar "head"

The polar "head" is soluble in water; the nonpolar "tail" is insoluble in water. The monolayer is composed of stearic acid molecules with their polar "heads" dissolved in the water and their nonpolar "tails" repelled from the surface of the water.

The **surface area** refers to the area occupied by all the molecules in the monolayer. The surface area of the monolayer corresponds to the diameter of the water in the watchglass (see Figure 11.2).

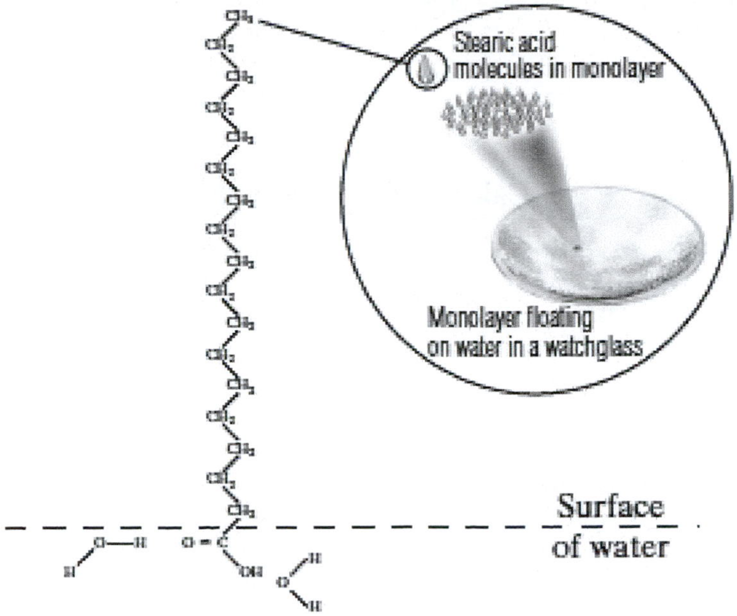

Figure 11.2 Enlarged View of Stearic Acid in Water The polar –COOH "head" dissolves in water, but the nonpolar "tail" of the molecule does not. The stearic acid molecules cover the entire surface of the water in the watchglass.

Example Exercise 11.1 • Calculating Molecules in the Monolayer

Drops of stearic acid solution on water give a monolayer with a diameter of 14.5 cm. Calculate the number of stearic acid molecules in the monolayer, assuming each molecule occupies an area of 0.21 nm^2.

Solution: First, we must find the surface area of the monolayer. We can calculate the area of a circle using the formula πr^2. Since the radius (r) is half the diameter (d), we can write the formula for the circular monolayer as:

$$\frac{\pi d^2}{4} = \text{surface area}$$

$$\frac{(3.14)(14.5\ \text{cm})^2}{4} = 165\ \text{cm}^2$$

Second, we can find the number of molecules by dividing the surface area of the monolayer (165 cm^2) by the area of a single molecule (0.21 nm^2). Since the areas are expressed in different units, we will convert cm^2 to nm^2.

$$\frac{165\ \text{cm}^2}{\text{monolayer}} \times \left(\frac{1\ \text{m}}{1 \times 10^2\ \text{cm}}\right)^2 \times \left(\frac{1 \times 10^9\ \text{nm}}{1\ \text{m}}\right)^2 \times \frac{1\ \text{molecule}}{0.21\ \text{nm}^2} = \frac{\text{molecules}}{\text{monolayer}}$$

$$\frac{165\ \cancel{\text{cm}^2}}{\text{monolayer}} \times \frac{1\ \cancel{\text{m}^2}}{1 \times 10^4\ \cancel{\text{cm}^2}} \times \frac{1 \times 10^{18}\ \cancel{\text{nm}^2}}{1\ \cancel{\text{m}^2}} \times \frac{1\ \text{molecule}}{0.21\ \cancel{\text{nm}^2}} = \frac{\text{molecules}}{\text{monolayer}}$$

$$= 7.9 \times 10^{16}\ \text{molecules/monolayer}$$

Moles of Stearic Acid in the Monolayer

Next, we can calculate the number of moles of stearic acid in the monolayer. Let's continue with the determination of Avogadro's number.

Example Exercise 11.2 • Calculating Moles in the Monolayer

A calibrated pipet (65 drops/mL) delivered 11 drops of stearic acid solution onto the monolayer. If the concentration of the stearic acid solution is 1.4×10^{-4} g/mL, how many moles of stearic acid (284 g/mol) are in the monolayer?

Solution: The moles of stearic acid in the monolayer are calculated as follows:

$$\frac{11 \text{ \sout{drops}}}{\text{monolayer}} \times \frac{1 \text{ \sout{mL solution}}}{65 \text{ \sout{drops}}} \times \frac{1.4 \times 10^{-4} \text{ \sout{g}}}{1 \text{ \sout{mL solution}}} \times \frac{1 \text{ mol}}{284 \text{ \sout{g}}} = \frac{\text{mol}}{\text{monolayer}}$$

$$= 8.3 \times 10^{-8} \text{ mol/monolayer}$$

In this example, the monolayer contains 8.3×10^{-8} moles of stearic acid.

Experimental Value for Avogadro's Number

Finally, we can calculate a value for Avogadro's number by comparing the molecules of stearic acid in the monolayer to the moles of stearic acid in the monolayer.

Example Exercise 11.3 • Determining Avogadro's Number

If the monolayer contains 7.9×10^{16} molecules of stearic acid and 8.3×10^{-8} moles, what is the experimental value for Avogadro's number?

Solution: The experimental value for Avogadro's number is simply the ratio of the number of stearic acid molecules per mole:

$$\text{Avogadro's number} = \frac{7.9 \times 10^{16} \text{ molecules}}{8.3 \times 10^{-8} \text{ mol}}$$

$$= 9.5 \times 10^{23} \text{ molecules/mol}$$

The experimental value for Avogadro's number is 9.5×10^{23}. This value is typical for the experiment, which tends to give somewhat high results.

EQUIPMENT and CHEMICALS

- dropper pipet (or 15 cm of 6-mm glass tubing)
- dropper pipet bulb
- 13 x 100 mm test tube & test tube rack
- 15 cm watchglass
- wash bottle with distilled water
- metric ruler
- stearic acid solution (0.00012–0.00015 g/mL of hexane is recommended)
- 10-mL graduated cylinder

PROCEDURE

A. Calibrating a Dropper Pipet

1. Cut a 15-cm length of 6-mm glass tubing. Heat the tubing and draw it into a fine capillary tip.

   Note: A commercial dropper pipet may be substituted; however, for best results the dropper calibration should exceed 50 drops/mL.

2. Obtain about 3 mL of stearic acid solution in a test tube. Calibrate the dropper pipet by adding drops of stearic acid solution into a 10-mL graduated cylinder. Hold the dropper at a 45° angle, and deliver the drops at a rate of one per second. Record the number of drops to reach the 1 mL mark.

   RECYCLE Chemical Waste

3. Repeat the dropwise calibration procedure twice and find the average number of drops for the three trials.

B. Calculating Molecules in the Monolayer

1. Measure the diameter of a large watchglass to 0.1 cm. Record the diameter, and calculate the surface area of the monolayer; that is, the surface area of the watchglass.

   Note: The diameter of a stearic acid monolayer corresponds to the surface area of the watchglass filled with water.

2. Calculate the number of molecules that can occupy a monolayer assuming the area of a stearic acid molecule is 0.21 nm^2.

C. Determining Avogadro's Number

1. Clean the watchglass carefully with soap and water. Rinse the watchglass thoroughly with distilled water, and do not touch the inside concave surface.

2. Place the convex side of the watchglass on a paper towel on the lab bench. Fill the watchglass completely with distilled water from a wash bottle.

RECYCLE Chemical Waste

3. Record the concentration of the stearic acid solution. Hold the dropper pipet at a 45° angle, and slowly deliver drops of solution onto the center of the water surface. Continue adding drops of solution until a clear lens persists for at least 30 seconds.

   Note: A clear lens lasting several seconds indicates a monolayer of molecules has formed across the entire surface of the water.

4. In the Data Table, record the number of drops required to form the monolayer.

5. Find the moles of stearic acid in the monolayer, and determine the experimental value for Avogadro's number.

6. Thoroughly clean and rinse the watchglass and perform a second trial.

7. Thoroughly clean and rinse the watchglass and perform a third trial.

EXPERIMENT 11 NAME ______________________

DATE ______________ SECTION ______________________

PRELABORATORY ASSIGNMENT*

1. Provide the key term that corresponds to each of the following definitions.

________ (a) 6.02×10^{23} particles; for example, 6.02×10^{23} stearic acid molecules

________ (b) a thin film layer of organic molecules on the surface of water

________ (c) the region occupied by a single layer of organic molecules floating on water

________ (d) the portion of a long organic molecule having polar bonds

________ (e) the portion of a long organic molecule having nonpolar bonds

________ (f) the mass of 1 mole of any substance expressed in grams

________ (g) the amount of substance that contains Avogadro's number of particles

Key Terms: Avogadro's number (*N*), molar mass (MM), mole (mol), monolayer, nonpolar "tail,"polar "head," surface area

2. Circle the portion of the stearic acid molecule shown below that dissolves in water.

CH_3-CH_2-CH_2-CH_2-CH_2-CH_2-CH_2-CH_2-CH_2-CH_2-CH_2-CH_2-CH_2-CH_2-CH_2-CH_2-CH_2–COOH

3. What is observed after the hexane solvent evaporates?

4. When do you stop adding drops of stearic acid solution to the watchglass?

5. What are major sources of experimental error?

* *Answers in Appendix J*

6. A calibrated pipet (75 drops/mL) delivers 12 drops of stearic acid solution (1.5×10^{-4} g/mL) and produces a monolayer with a diameter of 12.0 cm. Calculate:

   (a) the surface area of the monolayer (refer to Example Exercise 11.1).

   (b) the number of stearic acid molecules in the monolayer assuming each molecule occupies an area of 0.21 nm^2 (refer to Example Exercise 11.1).

   (c) the moles of stearic acid (284 g/mol) in the monolayer (refer to Example Exercise 11.2).

   (d) the experimental value for Avogadro's number (refer to Example Exercise 11.3).

7. What safety precautions must be observed in this experiment?

EXPERIMENT 11 NAME ____________________

DATE ____________ SECTION ____________________

DATA TABLE

A. Calibrating a Dropper Pipet

number of drops per milliliter ______ ______ ______

Average value ______ drops/mL

RECYCLE Chemical Waste

B. Calculating Molecules in the Monolayer

diameter of monolayer *(diameter of watchglass)* ______ cm

area of one stearic acid molecule 0.21 nm^2

Show the calculation for the surface area of the monolayer (see Example Exercise 11.1).

surface area of monolayer ______ cm^2

Show the calculation for the number of stearic acid molecules in the monolayer.

molecules in monolayer ______ molecules

RECYCLE
Chemical
Waste

C. Determining Avogadro's Number

number of drops in monolayer ________ ________ ________

concentration of stearic acid solution ________ g/mL
(see Instructor)

Show the calculation for the moles of stearic acid in the monolayer for trial 1 (see Example Exercise 11.2).

moles in monolayer ________ mol ________ mol ________ mol

Show the calculation for the number of molecules of stearic acid per mole for trial 1 (see Example Exercise 11.3).

Avogadro's number *(N)* ________ ________ ________

Average value for *N* ________ molecules/mol

EXPERIMENT 11 NAME ______________________

DATE ______________ SECTION ______________________

POSTLABORATORY ASSIGNMENT

1. An oleic acid, $C_{17}H_{33}COOH$ (282 g/mol), solution is added to water in a watchglass until a monolayer forms. Assume that there are no spaces between molecules in the monolayer and that each oleic acid molecule occupies an area of 0.25 nm^2. If the concentration of the oleic acid solution is 0.00012 g/mL, what is the experimental value of Avogadro's number?

dropper pipet calibration	=	65 drops/mL
number of drops of oleic acid in the monolayer	=	16 drops
diameter of monolayer	=	14.5 cm

N = ______________

2. Calculate the mass of carbon in a 5-carat diamond that contains 5.00 x 10^{22} atoms of carbon.

____

3. Calculate the mass of water that contains 10,000,000,000,000,000,000,000 molecules.

____

4. Calculate the number of O_3 molecules in 1.00 L of ozone gas at STP.

____

5. Calculate the number of O_3 molecules in 1.00 g of ozone gas.

____

6. Calculate the mass of one molecule of ozone, O_3, expressed in grams.

____

7. (optional) A student found the mass of a 250-mL beaker to be 110.546 g. After touching the beaker and leaving moisture fingerprints, the mass of the beaker was found to be 110.547 g. How many water molecules were transferred to the beaker by fingerprints?

____

EXPERIMENT

12

Empirical Formulas of Compounds

OBJECTIVES

- To determine the empirical formula for magnesium oxide.
- To determine the empirical formula for copper sulfide.
- To gain practical experience in developing techniques using a crucible.

DISCUSSION

During the late 1700s, chemists experimented with elements to see how they reacted to form compounds. In particular, they were interested in the reactions of metals as they combined with oxygen in the air. By measuring the mass of a metal before reaction and the mass of the metal oxide after reaction, chemists were able to determine the formulas of metal oxide compounds.

The simplest whole number ratio of atoms in a compound is referred to as the **empirical formula**. Mendeleev placed an element in a particular group in the periodic table based on the empirical formula of its oxide. For example, he placed magnesium, calcium, strontium, and barium in Group IIA/2 because they react with oxygen to give similar empirical formulas; that is, MgO, CaO, SrO, and BaO.

Since transition metals can combine with nonmetals in different ratios, we cannot always predict the empirical formulas of transition metal compounds. For example, iron can combine with oxygen to form either iron(II) oxide, FeO, or iron(III) oxide, Fe_2O_3. The following example exercises illustrate the calculation of empirical formulas.

Example Exercise 12.1 • Determining an Empirical Formula

A 0.279-g sample of iron is heated and allowed to react with oxygen from the air. If the product has a mass of 0.400 g, what is the empirical formula of the iron oxide?

Solution: The empirical formula is experimentally determined from the moles of each reactant. The moles of iron are calculated as follows.

$$0.279\ \cancel{\text{g Fe}} \times \frac{1\ \text{mol Fe}}{55.85\ \cancel{\text{g Fe}}} = 0.00500\ \text{mol Fe}$$

The mass of oxygen that reacted is 0.400 g product – 0.279 g iron = 0.121 g. We can calculate the moles of oxygen as follows.

$$0.121\ \cancel{\text{g O}} \times \frac{1\ \text{mol O}}{16.00\ \cancel{\text{g O}}} = 0.00756\ \text{mol O}$$

The mole ratio of the elements in iron oxide is $Fe_{0.00500}O_{0.00756}$, and we can divide by 0.00500 to find the simplest whole number ratio.

$$Fe_{\frac{0.00500}{0.00500}}\,O_{\frac{0.00756}{0.00500}} = Fe_{1.00}O_{1.51}$$

If we double the mole ratio, we obtain $Fe_2O_{3.02}$. The slight deviation from a whole number ratio is due to experimental error. Thus, the empirical formula is Fe_2O_3, and we name the compound iron(III) oxide, or ferric oxide.

Example Exercise 12.2 • Determining an Empirical Formula

A 0.331-g sample of iron is placed in a crucible and covered with powdered sulfur. The crucible is heated until all the excess sulfur is driven off. If the product weighs 0.522 g, what is the empirical formula of the iron sulfide?

Solution: First, we can calculate the moles of iron in the product.

$$0.331\ \cancel{\text{g Fe}} \times \frac{1\ \text{mol Fe}}{55.85\ \cancel{\text{g Fe}}} = 0.00593\ \text{mol Fe}$$

The mass of sulfur that reacted is 0.522 g product – 0.331 g iron = 0.191 g. Second, we can calculate the moles of sulfur as follows.

$$0.191\ \cancel{\text{g S}} \times \frac{1\ \text{mol S}}{32.07\ \cancel{\text{g S}}} = 0.00596\ \text{mol S}$$

The mole ratio of the elements in iron sulfide is $Fe_{0.00592}S_{0.00596}$, and we divide by 0.00592 to find the simplest whole number ratio.

$$Fe_{\frac{0.00593}{0.00593}}\,S_{\frac{0.00596}{0.00593}} = Fe_{1.00}S_{1.01}$$

The slight deviation from a whole number ratio is due to experimental error. Thus, the empirical formula for the product is FeS, and we name the compound iron(II) sulfide, or ferrous sulfide.

In this experiment, you will ignite magnesium ribbon in a crucible and convert the metal to an oxide product. The second part of the experiment involves the conversion of copper to copper sulfide. Since copper can form either copper(I) sulfide or copper(II) sulfide, the empirical formula is unknown and cannot be predicted. Figure 12.1 illustrates the experimental equipment.

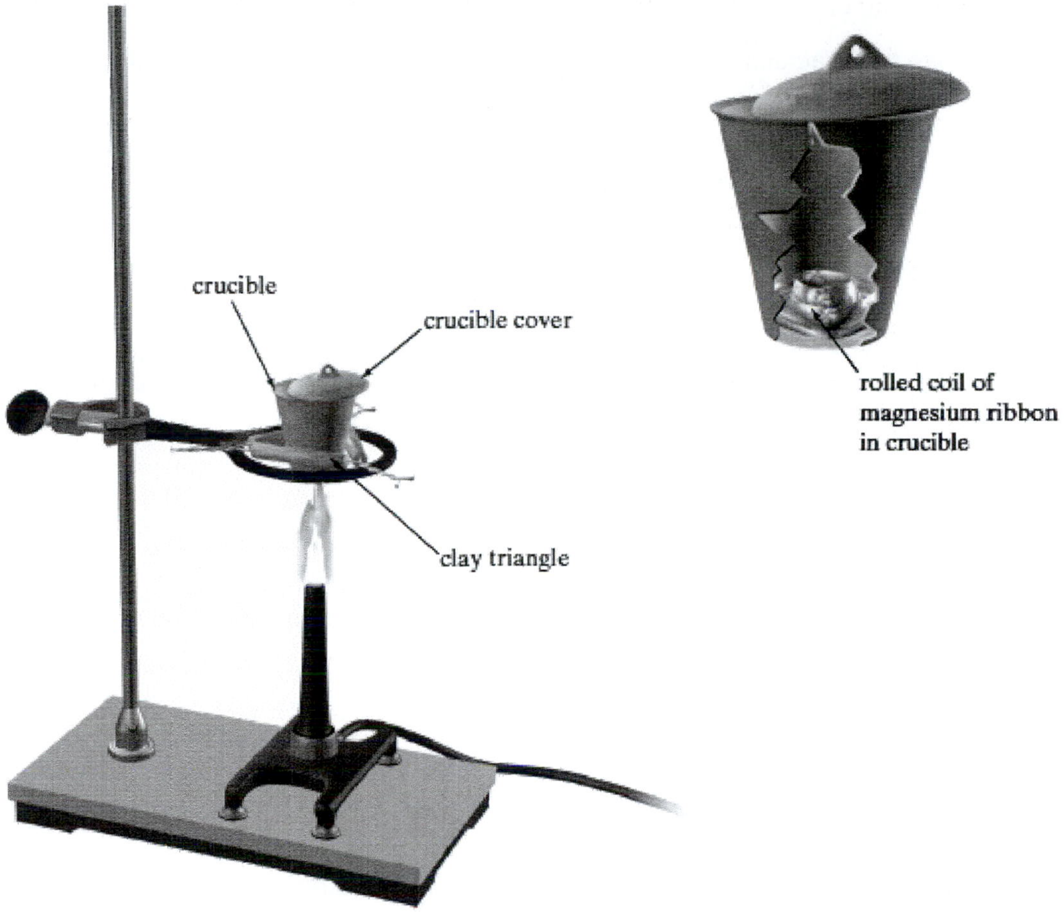

Figure 12.1 Empirical Formula Apparatus A crucible and cover are placed in a clay triangle on a ring stand, and heated until red hot.

EQUIPMENT and CHEMICALS

- crucible & cover
- clay triangle
- crucible tongs
- ring stand & ring

- magnesium, Mg ribbon
- copper, #18 gauge Cu wire
- sulfur, S powder

PROCEDURE

A. Empirical Formula of Magnesium Oxide

1. Support a crucible and cover with a clay triangle, and place on a ring stand. Fire the crucible and cover to red heat using the tip of the flame from a laboratory burner.
2. Remove the heat, and allow the crucible and cover to cool for 10 minutes. Weigh the crucible and cover.
3. Cut a 25-cm strip of magnesium ribbon, and roll the metal into a flat coil. Place the coil of magnesium at the bottom of the crucible. Reweigh the crucible, cover, and metal.
4. With the cover off, fire the crucible and magnesium to red heat. When the metal sparks and begins to smoke, immediately remove the burner and place the cover on the crucible using crucible tongs.
5. After the smoke has ceased, continue to heat the crucible and cover until the metal is completely converted to a gray-white residue. The progress of the reaction can be checked by lifting the cover with crucible tongs.
6. When the metal no longer sparks, turn off the burner and allow the crucible to cool for 10 minutes. Using a dropper pipet, add drops of distilled water until the gray-white residue no longer fizzes.

RECYCLE Chemical Waste

7. Cover the crucible, and heat for 5 minutes. Turn off the burner, and allow the crucible to cool for 10 minutes. Weigh the crucible and cover containing the magnesium oxide.
8. Clean the crucible, and repeat the procedure.
9. Calculate the empirical formula for each trial.

B. Empirical Formula of Copper Sulfide

Caution: This procedure requires a vented fume hood, as burning sulfur produces pungent sulfur dioxide gas.

1. Support a crucible and cover with a clay triangle, and place on a ring stand. Fire the crucible and cover to red heat.
2. Remove the heat, and allow the crucible and cover to cool for 10 minutes. Weigh the crucible and cover.
3. Cut a 25-cm length of copper wire, and roll the wire into a coil. Place the coil of wire in the bottom of the crucible. Reweigh the crucible, cover, and copper wire.
4. Cover the copper wire completely with powdered sulfur. Place the cover on the crucible, and gradually heat to red heat under a fume hood. Continue to heat for several minutes after the last trace of burning sulfur disappears.
5. Allow the crucible and contents to cool for 10 minutes. Weigh the crucible and cover containing the copper sulfide.
6. Clean the crucible, and repeat the procedure.
7. Calculate the empirical formula for each trial.

EXPERIMENT 12 NAME ______________________________

DATE ____________________ SECTION ______________________________

*PRELABORATORY ASSIGNMENT**

1. Provide the key term that corresponds to each of the following definitions.

___________ (a) the chemical formula of a compound that expresses the simplest whole number ratio of atoms of each element in a molecule, or ions in a formula unit

___________ (b) the chemical formula of a compound that expresses the actual number of atoms of each element in a molecule

___________ (c) heating a crucible until it glows red

___________ (d) a repeated process of heating, cooling, and weighing until the mass readings for an object are constant, or agree closely

___________ (e) a procedure for obtaining the mass of a sample indirectly by first weighing a container and then weighing the container with the sample

___________ (f) the mass of 1 mole of any substance expressed in grams

___________ (g) the amount of substance that contains Avogadro's number of particles

Key Terms: empirical formula, firing to red heat, heating to constant weight, molar mass (MM), mole (mol), molecular formula, weighing by difference

2. Why are the empty crucible and cover fired to red heat?

3. How critical are the suggested times for heating and cooling?

4. Why is distilled water added to the crucible after igniting the magnesium metal?

5. How can you tell when the magnesium metal has reacted completely?

** Answers in Appendix J*

6. How can you tell when the copper wire has reacted completely?

7. A sample of magnesium ribbon is ignited in a crucible to form magnesium oxide. Refer to Example Exercise 12.1 and determine the empirical formula of magnesium oxide from the following data:

mass of crucible and cover + magnesium metal	33.741 g
mass of crucible and cover	33.500 g
mass of crucible and cover + magnesium oxide	33.899 g

8. What are major sources of experimental error?

9. What safety precautions must be observed in this experiment?

EXPERIMENT 12 NAME ______________________

DATE ______________ SECTION ______________________

DATA TABLE

A. Empirical Formula of Magnesium Oxide

mass of crucible and cover + magnesium metal *(before heating)*	________ g	________ g
mass of crucible and cover	________ g	________ g
mass of magnesium metal	________ g	________ g
mass of crucible and cover + magnesium oxide (*after heating*)	________ g	________ g
mass of combined oxygen *(after heating – before heating)*	________ g	________ g

RECYCLE Chemical Waste

Show the calculation of the empirical formula for trial 1 (see Example Exercise 12.1).

Empirical formula of magnesium oxide ________ ________

B. Empirical Formula of Copper Sulfide

mass of crucible and cover + copper wire *(before heating)*	_______ g	_______ g
mass of crucible and cover	_______ g	_______ g
mass of copper wire	_______ g	_______ g
mass of crucible and cover + copper sulfide *(after heating)*	_______ g	_______ g
mass of crucible and cover + copper sulfide *(optional second heating)*	_______ g	_______ g
mass of combined sulfur *(after heating – before heating)*	_______ g	_______ g

RECYCLE
Chemical
Waste

Show the calculation of the empirical formula for trial 1 (see Example Exercise 12.2).

Empirical formula of copper sulfide	_______	_______

EXPERIMENT 12 NAME ______________________

DATE ______________ SECTION ______________________

POSTLABORATORY ASSIGNMENT

1. Refer to the periodic table and predict the empirical formula for each of the following oxides given the formula of aluminum oxide, Al_2O_3.

   (a) $B_?O_?$ ________ (b) $Ga_?O_?$ ________

   (c) $In_?O_?$ ________ (d) $Tl_?O_?$ ________

2. Refer to the periodic table and predict the empirical formula for each of the following halides given the formula of magnesium chloride, $MgCl_2$.

   (a) magnesium fluoride ________ (b) calcium chloride ________

   (c) strontium bromide ________ (d) barium iodide ________

3. A 1.000-g sample of lead shot reacted with oxygen to give 1.077 g of product. Calculate the empirical formula of the lead oxide.

   ____

4. A 0.500-g sample of tin foil reacted with oxygen to give 0.635 g of product. Calculate the empirical formula of the tin oxide.

   ____

5. A 1.000-g sample of cobalt metal reacted with sulfur powder to give 1.544 g of product. Calculate the empirical formula of the cobalt sulfide.

____

6. A 0.500-g sample of chromium metal reacted with sulfur powder to give 0.963 g of product. Calculate the empirical formula of the chromium sulfide.

____

7. (optional) Mercury chloride is a commercial fungicide. If the molar mass is 470 g/mol and the percent composition is 85.0% Hg and 15.0% Cl, what is the empirical and molecular formula of the fungicide?

Empirical formula ______________________

Molecular formula ______________________

Analysis of Alum

EXPERIMENT

13

OBJECTIVES

- To determine the percentage of water in alum hydrate.
- To determine the percentage of water in an unknown hydrate.
- To calculate the water of crystallization for an unknown hydrate.
- To develop the laboratory skills for analyzing a hydrate.

DISCUSSION

A **hydrate** is a compound having a fixed number of water molecules. The number of water molecules is referred to as the **water of crystallization**, or water of hydration. For example, barium chloride dihydrate, $BaCl_2 \cdot 2H_2O$, has two waters of crystallization; and alum hydrate, $KAl(SO_4)_2 \cdot 12H_2O$, has twelve waters of crystallization. When heated, a hydrate loses water and produces an **anhydrous compound**. Alum is a hydrate compound used in styptic pencils to stop minor bleeding, and decomposes to an anhydrous compound as follows:

$$\underset{\text{alum hydrate}}{KAl(SO_4)_2 \cdot 12H_2O(s)} \xrightarrow{\Delta} \underset{\text{anhydrous compound}}{KAl(SO_4)_2(s)} + \underset{\text{water of crystallization}}{12\,H_2O(g)}$$

The theoretical percentage of water in a hydrate is found by comparing the mass of the water of crystallization to the mass of the hydrate. This is accomplished by dividing the total mass of water by the molar mass of the hydrate. Example Exercise 13.1 illustrates this calculation.

Example Exercise 13.1 • Theoretical % H_2O in Alum

Calculate the theoretical percentage of water in alum hydrate, $KAl(SO_4)_2 \cdot 12H_2O$.

Solution: The molar mass of $KAl(SO_4)_2 \cdot 12H_2O$ is found as follows:

K:	1	x	39.10 g	=	39.10 g
Al:	1	x	26.98 g	=	26.98 g
SO_4:	2	x	96.07 g	=	192.14 g
H_2O:	12	x	18.02 g	=	216.24 g
					474.46 g

The theoretical percentage of water is found by dividing the total mass of water (12 x 18.02 g) by the molar mass of the hydrate (474.46 g).

$$\frac{12 \times 18.02 \text{ g}}{474.46 \text{ g}} \times 100\% = 45.58\%\ H_2O$$

Percentage of Water in Alum Hydrate

Experimentally, the amount of water is found from **weighing by difference**; that is, the difference in mass before and after heating the sample. For example, an alum hydrate sample is decomposed to give the following data:

mass of beaker and watchglass + alum hydrate	102.636 g
mass of beaker and watchglass	101.486 g
mass of beaker and watchglass + anhydrous compound	102.113 g

The mass of alum hydrate is: 102.636 g – 101.486 g = 1.150 g. The mass of water is the difference before and after heating the sample: 102.636 g – 102.113 g = 0.523 g. The calculation for the experimental percentage of water is illustrated in Example Exercise 13.2.

Example Exercise 13.2 • Experimental % H_2O in Alum

A 1.150-g sample of alum hydrate decomposes with heat and loses 0.523 g of water. Calculate the experimental percentage of water in alum sample.

Solution: The experimental percentage of water is simply

$$\frac{\text{mass of water}}{\text{mass of hydrate}} \times 100\% = \%\text{ water}$$

$$\frac{0.523 \text{ g}}{1.150 \text{ g}} \times 100\% = 45.5\%\text{ water}$$

In this experiment, we will heat a sample of alum hydrate to determine the percentage of water (Figure 13.1). After gaining experience analyzing the known alum sample, you will analyze an unknown hydrate.

Water of Crystallization in an Unknown Hydrate

After analyzing an alum sample, you will analyze an unknown hydrate for the percentage of water using the same procedure. You will also determine the water of crystallization for the unknown hydrate; that is, you will determine the value of *X* in an unknown hydrate, $AC \cdot XH_2O$.

Example Exercise 13.3 • Water of Crystallization for Unknown Hydrate

Calculate the water of crystallization for an unknown hydrate, $AC \cdot XH_2O$, that is found to contain 30.6% water. The molar mass of the anhydrous compound (AC) is 245 g/mol.

Solution: If the amount of water in the unknown hydrate is 30.6%, the anhydrous compound must be 69.4% (100% – 30.6% = 69.4%). If we assume a 100.0-g sample, the mass of water is 30.6 g and the anhydrous compound is 69.4 g. We can calculate the moles of water and anhydrous compound as follows.

$$30.6 \cancel{\text{g } H_2O} \times \frac{1 \text{ mol } H_2O}{18.02 \cancel{\text{g } H_2O}} = 1.70 \text{ mol } H_2O$$

$$69.4 \cancel{\text{g AC}} \times \frac{1 \text{ mol AC}}{245 \cancel{\text{g AC}}} = 0.283 \text{ mol AC}$$

To find the water of crystallization we simply divide the moles of water by the moles of anhydrous compound. Thus,

$$\frac{1.70 \text{ mol } H_2O}{0.283 \text{ mol AC}} = 6.01 \approx 6$$

Since the water of crystallization must be a whole number, we round off 6.01 to the nearest whole number (6). The formula for the hydrate is $AC \cdot 6H_2O$.

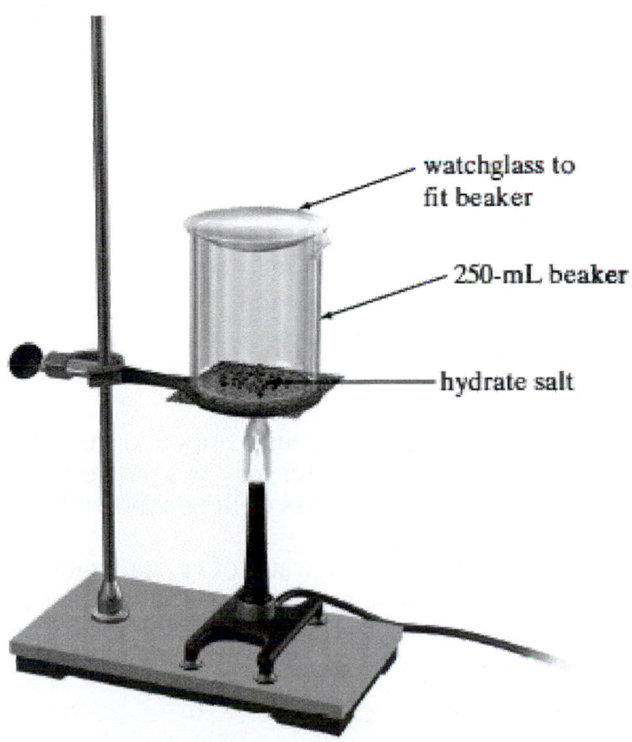

Figure 13.1 Decomposition Apparatus Gently heat a hydrate salt in the beaker to release steam and avoid spattering.

EQUIPMENT and CHEMICALS

- wire gauze
- 250-mL beaker
- watchglass
- ring stand & ring
- alum, $KAl(SO_4)_2 \cdot 12H_2O$
- unknown hydrate samples

PROCEDURE

A. Percentage of Water in Alum Hydrate

1. Weigh a clean, dry, 250-mL beaker covered with a watchglass. Add about 0.8–1.2 g of alum hydrate into the beaker, and reweigh.

2. Support the beaker and watchglass on a ring stand using a wire gauze (Figure 13.1). Gently heat the hydrate and observe moisture on the sides of the beaker and bottom of the watchglass. Continue heating until all the moisture is evaporated. When the hydrate is completely decomposed, it will change from crystalline to powder.

   Note: If the watchglass is not completely dry, hold it carefully with crucible tongs over a low burner flame until no moisture remains.

3. Turn off the burner, and allow the beaker to cool for 10 minutes. Carefully transfer the beaker with the watchglass to the balance. Weigh the mass of the beaker, watchglass, and anhydrous compound.

RECYCLE Chemical Waste

4. Discard the decomposed alum and clean the beaker. Perform a second trial with alum hydrate. Calculate the percentage of water in the hydrate for each trial and the average value.

B. Percentage of Water in an Unknown Hydrate

1. Obtain an unknown hydrate from the Instructor, and record the unknown number.

2. Repeat steps 1–4 as in Procedure A, and report the average percentage of water in the unknown hydrate.

C. Water of Crystallization in an Unknown Hydrate

1. Given the molar mass of the anhydrous compound from the Instructor, calculate the water of crystallization. The Instructor may wish to verify the experimental percentage of water before giving the molar mass of the unknown hydrate.

   Note: It is not unusual for the water of crystallization value to differ by a few tenths from a whole number; for example, 2.1 or 6.7. However, the water of crystallization must be rounded to a whole number. In this example, 2.1 is rounded to 2 and 6.7 is rounded to 7.

EXPERIMENT 13 NAME ______________________________

DATE ____________________ SECTION ______________________________

*PRELABORATORY ASSIGNMENT**

1. Provide the key term that corresponds to each of the following definitions.

__________ (a) a substance that contains a specific number of water molecules attached to a formula unit in a crystalline compound

__________ (b) the number of water molecules bound to a formula unit in a hydrate

__________ (c) a compound that does not contain water

__________ (d) a procedure for obtaining the mass of a sample indirectly by first weighing a container and then weighing the container with the sample

__________ (e) a repeated process of heating, cooling, and weighing until the mass readings for an object are constant, or agree closely

__________ (f) the mass of 1 mole of any substance expressed in grams

__________ (g) the amount of substance that contains Avogadro's number of particles

Key Terms: anhydrous compound, hydrate, heating to constant weight, molar mass (MM), mole (mol), water of crystallization, weighing by difference

2. If you weigh a bag of microwave popcorn, before and after heating, does the popped corn weigh more or less than the unpopped kernels?

3. How can you tell when to stop heating a hydrate because it is decomposed completely?

4. How will the weighing be affected by placing a warm beaker on the balance?

5. What are major sources of experimental error?

---

* *Answers in Appendix J*

6. An alum hydrate sample was analyzed by decomposition and gave the following data:

mass of beaker and watchglass + alum hydrate	102.218 g
mass of beaker and watchglass	101.286 g
mass of beaker and watchglass + anhydrous compound	101.798 g

Refer to Example Exercise 13.2 and show the calculation for the percentage of water.

Does the experimental result agree with the theoretical value shown in Example Exercise 13.1?

7. An unknown hydrate ($AC \cdot X H_2O$) was found to contain 30.5% water. Assume the molar mass of the anhydrous compound (AC) is 164 g/mol. Refer to Example Exercise 13.3 and show the calculation for the water of crystallization and formula for the hydrate.

Water of crystallization ______ Formula of hydrate $AC \cdot$ ____ H_2O

8. What safety precautions must be observed in this experiment?

EXPERIMENT 13 NAME ______________________

DATE ______________ SECTION ______________________

DATA TABLE

A. Percentage of Water in Alum Hydrate

	Trial 1	Trial 2
mass of beaker and watchglass + alum hydrate *(before heating)*	________ g	________ g
mass of beaker and watchglass	________ g	________ g
mass of alum hydrate	________ g	________ g
mass of beaker and watchglass + anhydrous compound *(after heating)*	________ g	________ g
mass of water *(before heating – after heating)*	________ g	________ g

RECYCLE
Chemical Waste

Show the calculation for the percentage of water for trial 1 (see Example Exercise 13.2).

Percentage of water in $KAl(SO_4)_2 \cdot 12H_2O$	________ %	________ %

Average percentage of water ____________ %

B. Percentage of Water in an Unknown Hydrate **UNKNOWN #** ______

mass of beaker and watchglass + unknown hydrate *(before heating)*	________ g	________ g
mass of beaker and watchglass	________ g	________ g
mass of unknown hydrate	________ g	________ g
mass of beaker and watchglass + anhydrous compound *(after heating)*	________ g	________ g
mass of water *(before heating – after heating)*	________ g	________ g

RECYCLE Chemical Waste

Show the calculation for the percentage of water for trial 1 (see Example Exercise 13.2).

Percentage of water in the unknown hydrate	________ %	________ %
Average percentage of water	________ %	

C. Water of Crystallization in an Unknown Hydrate

molar mass of anhydrous compound (AC) *(see Instructor)*	________ g/mol
percentage of water (see Procedure B)	________ %
percentage of anhydrous compound (AC)	________ %

Show the calculation for the water of crystallization (see Example Exercise 13.3).

Water of crystallization ______ Formula of hydrate AC • ____ H_2O

EXPERIMENT 13 NAME ______________________

DATE ______________ SECTION ______________________

POSTLABORATORY ASSIGNMENT

1. Calculate the theoretical percentage of water for the following hydrates.

   (a) sodium carbonate hexahydrate, $Na_2CO_3 \cdot 6H_2O$

   ____

   (b) sodium carbonate decahydrate, $Na_2CO_3 \cdot 10H_2O$

   ____

2. An unknown hydrate, $AC \cdot XH_2O$, has a mass of 1.000 g before heating, and a mass of 0.738 g after heating. What is the experimental percentage of water in the hydrate?

   ____

3. If the anhydrous compound (AC) in the preceding problem has a molar mass of 101 g/mol, what is the water of crystallization (X) and the formula for the hydrate?

   Water of crystallization ______ Formula of hydrate $AC \cdot$ ____ H_2O

4. A hydrate of nickel(II) chloride, $NiCl_2 \cdot XH_2O$, decomposes to produce 21.8% water. Calculate the water of crystallization (X), and write the formula for the hydrate.

Water of crystallization ______ Formula of hydrate $NiCl_2 \cdot$ ____ H_2O

5. A different hydrate of nickel(II) chloride, $NiCl_2 \cdot XH_2O$, decomposes to produce 45.5% water. Calculate the water of crystallization (X), and write the formula for the hydrate.

Water of crystallization ______ Formula of hydrate $NiCl_2 \cdot$ ____ H_2O

6. (optional) A blue turquoise is an example of a hydrate mineral containing copper; the chemical formula is $CuAl_6(PO_4)_4(OH)_8 \cdot 4H_2O$. What is the percent water in this semiprecious stone?

____

EXPERIMENT

14

Decomposing Baking Soda

OBJECTIVES

- To determine the percent yield of sodium carbonate from a decomposition reaction.
- To determine the percentage of sodium hydrogen carbonate in an unknown mixture.
- To gain proficiency in decomposing a compound and collecting a gas over water.

DISCUSSION

When baking soda is heated, sodium hydrogen carbonate, $NaHCO_3$, decomposes into solid sodium carbonate, while releasing steam and carbon dioxide gas. The equation for the reaction is

$$2\ NaHCO_3(s) \xrightarrow{\Delta} Na_2CO_3(s) + H_2O(g) + CO_2(g)$$

Notice that the reaction releases H_2O and CO_2 as gases but Na_2CO_3 remains a solid. If we *weigh* the mass of solid Na_2CO_3 produced in an experiment, the mass is referred to as the **actual yield**. Conversely, if we *calculate* the mass of Na_2CO_3 according to the balanced chemical equation, the mass is referred to as the **theoretical yield**.

The **percent yield** from a chemical reaction is an expression for the amount of actual yield compared to the theoretical yield. While some experimental errors give high results, other errors give low results. Thus, the percent yield can be greater than—or less than—100%.

Percent Yield of Sodium Carbonate from Baking Soda

Example Exercise 14.1 • % Yield of Na_2CO_3 from Baking Soda

A 1.654-g sample of pure baking soda, $NaHCO_3$, decomposes to produce 1.028 g of solid sodium carbonate. Calculate the theoretical yield and percent yield of Na_2CO_3.

Solution: According to the balanced equation, 2 mol $NaHCO_3$ (84.01 g/mol) produce 1 mol Na_2CO_3 (105.99 g/mol). We can find the theoretical yield as follows:

$$1.654 \text{ g } NaHCO_3 \times \frac{1 \text{ mol } NaHCO_3}{84.01 \text{ g } NaHCO_3} \times \frac{1 \text{ mol } Na_2CO_3}{2 \text{ mol } NaHCO_3} \times \frac{105.99 \text{ g } Na_2CO_3}{1 \text{ mol } Na_2CO_3} = 1.043 \text{ g } Na_2CO_3$$

Since the actual yield of Na_2CO_3 is 1.028 g, the percent yield is

$$\frac{\text{actual yield}}{\text{theoretical yield}} \times 100\% = \% \text{ yield}$$

$$\frac{1.028 \text{ g}}{1.043 \text{ g}} \times 100\% = 98.56\%$$

Percentage of Sodium Hydrogen Carbonate in an Unknown Mixture

An unknown mixture containing baking soda is decomposed using heat. The following example exercise illustrates the calculation for the percentage of baking soda in the mixture.

Example Exercise 14.2 • % $NaHCO_3$ in an Unknown Mixture

A 1.675-g unknown mixture containing baking soda is decomposed with heat. If the mass loss is 0.318 g, what is the percentage of baking soda, $NaHCO_3$, in the unknown mixture?

Solution: In this example, the mass loss corresponds to both the mass of water vapor and carbon dioxide gas. To simplify the calculation, we will combine $H_2O + CO_2$ into H_2CO_3 (62.03 g/mol) and rewrite the chemical equation.

$$2\ NaHCO_3(s) \xrightarrow{\Delta} Na_2CO_3(s) + H_2CO_3(g)$$

We can relate the H_2CO_3 mass loss to the mass of $NaHCO_3$ as follows:

$$0.318 \text{ g } H_2CO_3 \times \frac{1 \text{ mol } H_2CO_3}{62.03 \text{ g } H_2CO_3} \times \frac{2 \text{ mol } NaHCO_3}{1 \text{ mol } H_2CO_3} \times \frac{84.01 \text{ g } NaHCO_3}{1 \text{ mol } NaHCO_3} = 0.861 \text{ g } NaHCO_3$$

If the sample mixture has a mass of 1.675 g, the percentage of $NaHCO_3$ is

$$\frac{\text{mass } NaHCO_3}{\text{mass sample}} \times 100\% = \% \ NaHCO_3$$

$$\frac{0.861 \text{ g}}{1.675 \text{ g}} \times 100\% = 51.4\%$$

Figure 14.1 shows the experimental apparatus for decomposing baking soda. As the baking soda decomposes, carbon dioxide gas is produced. The carbon dioxide gas displaces water from the Florence flask into the 1000-mL beaker. When the decomposition is complete, no more carbon dioxide gas is released and the water level in the beaker remains constant.

Using a gas collection apparatus shown in 14.1 provides a visual learning experience for collecting a gas over water. That is, the carbon dioxide gas released from heating baking soda displaces water from the Florence flask into the beaker, If the apparatus is not readily available, this experiment can be performed by heating baking soda in a crucible and cover.

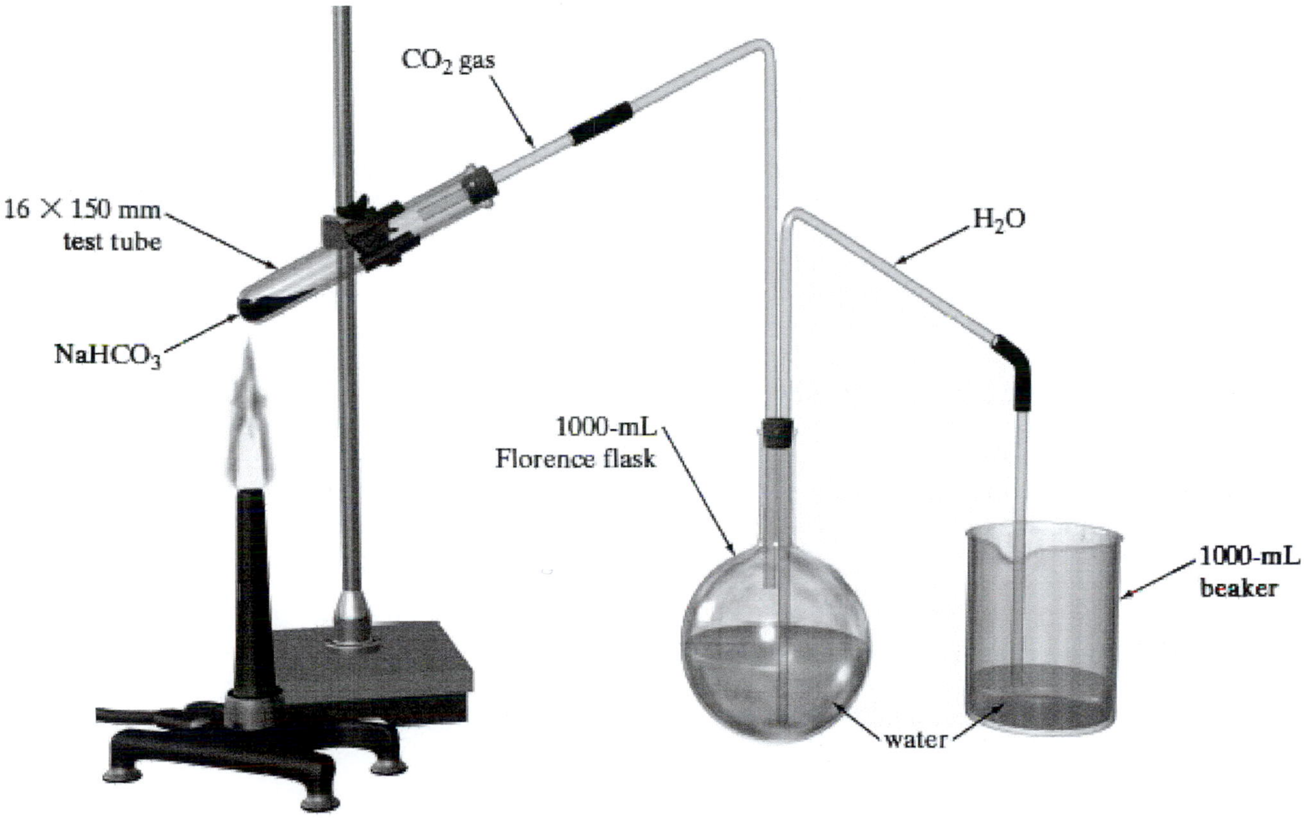

Figure 14.1 Decomposition Apparatus When the water level in the beaker remains constant, the decomposition of $NaHCO_3$ is complete.

EQUIPMENT and CHEMICALS

- gas collection apparatus (see Figure 14.1)
- 16 x 150 mm test tube
- 1000-mL Florence flask
- 1000-mL beaker
- sodium hydrogen carbonate, baking soda, solid $NaHCO_3$
- unknown baking soda mixture, 50–90% $NaHCO_3$

PROCEDURE

A. Percent Yield of Na_2CO_3 from Baking Soda

1. Weigh a 16 x 150 mm *dry* test tube on the balance, and record the mass. Add 1–2 g of baking soda, $NaHCO_3$, and reweigh.

2. Set up the apparatus as shown in Figure 14.1. Fill the Florence flask to the neck with tap water, and insert the gas collection apparatus. Insert the small rubber stopper into the test tube as shown.

3. Gently heat the test tube and baking soda. Observe water being displaced into the beaker as carbon dioxide gas is produced. As the water level in the beaker increases, continue to heat the test tube with a gentle flame. After the water level remains constant for a couple of minutes, discontinue heating and allow the test tube to cool for 10 minutes.

   Note: The decomposition of baking soda produces steam that may collect in the test tube. Any moisture in the test tube leads to serious weighing errors. If there appears to be moisture in the test tube, remove the utility clamp from the ring stand and carefully heat the open test tube over a low flame until no trace of moisture remains. Allow the test tube to cool for 10 minutes before weighing.

RECYCLE
Chemical
Waste

4. Weigh the test tube containing the sodium carbonate residue. The mass of Na_2CO_3 is found by subtracting the mass of the test tube from the test tube and residue.

5. Calculate the theoretical yield of sodium carbonate, Na_2CO_3, from the mass of pure baking soda that was heated. Find the percent yield of sodium carbonate.

B. Percentage of $NaHCO_3$ in an Unknown Mixture

1. Obtain an unknown sample containing baking soda. Record the unknown number in the Data Table.

2. Repeat steps 1–5 as in Procedure A; substitute an unknown baking soda mixture for pure baking soda.

3. Calculate the mass of baking soda, $NaHCO_3$, in the unknown sample from the mass loss. Find the percentage of baking soda in the unknown mixture.

EXPERIMENT 14 NAME ______________________________

DATE ____________________ SECTION ______________________________

*PRELABORATORY ASSIGNMENT**

1. Provide the key term that corresponds to each of the following definitions.

___________ (a) the relationship of quantities (i.e., mass of substance or volume of gas) in a chemical reaction according to the balanced chemical equation

___________ (b) the mass of 1 mole of any substance expressed in grams

___________ (c) a procedure for obtaining the mass of a sample by first weighing a container and then weighing the container with the sample

___________ (d) a technique for determining the volume of a gas by measuring the volume of water it displaces

___________ (e) the amount of product experimentally obtained from a reaction

___________ (f) the amount of product that is calculated from a given amount of reactant

___________ (g) the actual yield compared to the theoretical yield expressed as a percent

Key Terms: actual yield, molar mass (MM), percent yield, stoichiometry, theoretical yield, volume by displacement, weighing by difference

2. How do you tell when the baking soda sample is decomposed completely?

3. Is it possible to obtain a percent yield of Na_2CO_3 that is greater than 100%?

4. What are major sources of experimental error?

---

* *Answers in Appendix J*

5. A 1.555-g sample of baking soda decomposes with heat to produce 0.991 g Na_2CO_3. Refer to Example Exercise 14.1 and show the calculation for the theoretical yield of Na_2CO_3.

____

What is the percent yield of sodium carbonate, Na_2CO_3?

____

6. A 1.473-g unknown mixture with baking soda is heated and has a mass loss of 0.325 g. Refer to Example Exercise 14.2 and show the calculation for the percentage $NaHCO_3$ in the mixture.

____

7. What safety precautions must be observed in this experiment?

EXPERIMENT 14 NAME ____________________

DATE ____________ SECTION ____________________

DATA TABLE

A. Percent Yield of Na_2CO_3 from Baking Soda

	Trial 1	Trial 2
mass of test tube + $NaHCO_3$ *(before heating)*	________ g	________ g
mass of test tube	________ g	________ g
mass of $NaHCO_3$	________ g	________ g
mass of test tube + Na_2CO_3 *(after heating)*	________ g	________ g
mass of Na_2CO_3 *(actual yield)* *(after heating – test tube)*	________ g	________ g

RECYCLE Chemical Waste

Show the calculation for theoretical yield of Na_2CO_3 for trial 1 (see Example Exercise 14.1).

	Trial 1	Trial 2
mass of Na_2CO_3 *(theoretical yield)*	________ g	________ g

Show the calculation for percent yield of Na_2CO_3 for trial 1 (see Example Exercise 14.1).

	Trial 1	Trial 2
Percent Yield of Na_2CO_3	________ %	________ %

Average Percent Yield ________ %

B. Percentage of $NaHCO_3$ in an Unknown Mixture **UNKNOWN #** ______

mass of test tube + unknown mixture *(before heating)*	________ g	________ g
mass of test tube	________ g	________ g
mass of unknown mixture	________ g	________ g
mass of test tube + residue *(after heating)*	________ g	________ g
mass of H_2CO_3 ($H_2O + CO_2$) *(before heating – after heating)*	________ g	________ g

RECYCLE
Chemical
Waste

Show the calculation for the mass of $NaHCO_3$ in the unknown mixture for trial 1 (see Example Exercise 14.2).

mass of $NaHCO_3$ ________ g ________ g

Show the calculation for the percentage of $NaHCO_3$ in the unknown mixture for trial 1 (see Example Exercise 14.2).

Percentage of $NaHCO_3$ ________ % ________ %

Average percentage of $NaHCO_3$ ________ %

EXPERIMENT 14 NAME ______________________

DATE ______________ SECTION ______________________

POSTLABORATORY ASSIGNMENT

1. A 1.225-g sample of pure lithium hydrogen carbonate is decomposed by heating to produce 0.660 g lithium carbonate. Calculate the theoretical yield and percent yield of Li_2CO_3.

$$2\ LiHCO_3(s) \xrightarrow{\Delta} Li_2CO_3(s) + H_2O(g) + CO_2(g)$$

____

____

2. A 1.205-g sample mixture of lithium hydrogen carbonate is decomposed by heating. If the mass loss is 0.275 g, what is the percentage of $LiHCO_3$ in the unknown mixture?

$$2\ LiHCO_3(s) \xrightarrow{\Delta} Li_2CO_3(s) + H_2CO_3(g)$$

____

3. Lithium chlorate is decomposed with heat to give lithium chloride and oxygen gas. If 1.115 g of lithium chlorate is decomposed, how many milliliters of oxygen gas is released at STP?

$$2\ LiClO_3(s) \xrightarrow{\Delta} 2\ LiCl(s) + 3\ O_2(g)$$

____

4. Lithium metal reacts with water to give lithium hydroxide and hydrogen gas. If 75.5 mL of hydrogen gas is produced at STP, what is the mass of lithium metal that reacted?

$$2\ Li(s) + 2\ H_2O(l) \rightarrow 2\ LiOH(aq) + H_2(g)$$

____

5. (optional) The Solvay process is used to manufacture baking soda, $NaHCO_3$. In the process, CO_2, NH_3, H_2O, and NaCl react to produce baking soda. If 25.0 L CO_2 and 20.0 L NH_3 react at STP, with excess water and sodium chloride, what is the limiting reactant? Calculate the mass of baking soda produced.

$$CO_2(g) + NH_3(g) + H_2O(l) + NaCl(s) \rightarrow NaHCO_3(s) + NH_4Cl(aq)$$

Limiting Reactant ______________ Mass $NaHCO_3$ ______________

EXPERIMENT

15

Precipitating Calcium Phosphate

OBJECTIVES

- To determine the percent yield of calcium phosphate from a precipitation reaction.
- To determine the percentage of calcium chloride in an unknown mixture.
- To gain proficiency in transferring and filtering a precipitate.

DISCUSSION

In this experiment, we will use aqueous solutions to produce a precipitate of insoluble calcium phosphate, $Ca_3(PO_4)_2$. The equation for the reaction is

$$3\ CaCl_2(aq) + 2\ Na_3PO_4(aq) \rightarrow Ca_3(PO_4)_2(s) + 6\ NaCl(aq)$$

The precipitate will be collected in filter paper, which separates the insoluble particles from aqueous solution. The experimental mass of the precipitate is referred to as the **actual yield**. The calculated mass of the precipitate using the above equation is referred to as the **theoretical yield**.

The **percent yield** from a chemical reaction is an expression for the amount of actual yield compared to the theoretical yield. While some experimental errors lead to high results, other errors may give low results. Thus, the percent yield can be greater than—or less than—100%.

Percent Yield of Calcium Phosphate from Calcium Chloride

Example Exercise 15.1 • % Yield of $Ca_3(PO_4)_2$ from $CaCl_2$

A 0.555-g sample of calcium chloride gives a 0.505-g precipitate of calcium phosphate. Calculate the theoretical yield and percent yield of $Ca_3(PO_4)_2$.

Solution: According to the balanced chemical equation, 3 mol $CaCl_2$ (110.98 g/mol) produce 1 mol $Ca_3(PO_4)_2$ (310.18 g/mol) precipitate. We can calculate the theoretical mass of precipitate as follows:

$$0.555\ \cancel{g\ CaCl_2} \times \frac{1\ \cancel{mol\ CaCl_2}}{110.98\ \cancel{g\ CaCl_2}} \times \frac{1\ \cancel{mol\ Ca_3(PO_4)_2}}{3\ \cancel{mol\ CaCl_2}} \times \frac{310.18\ g\ Ca_3(PO_4)_2}{1\ \cancel{mol\ Ca_3(PO_4)_2}}$$

$$= 0.517\ g\ Ca_3(PO_4)_2$$

Since the actual yield of $Ca_3(PO_4)_2$ is 0.505 g, the percent yield is

$$\frac{\text{actual yield}}{\text{theoretical yield}} \times 100\% = \%\ \text{yield}$$

$$\frac{0.505\ \cancel{g}}{0.517\ \cancel{g}} \times 100\% = 97.7\%$$

Percentage of Calcium Chloride in an Unknown Mixture

When an unknown mixture containing calcium chloride reacts with sodium phosphate, the equation for the reaction is the same as above. In this calculation, however, we will relate the mass of calcium phosphate to the mass of calcium chloride in the original unknown mixture.

Example Exercise 15.2 • % $CaCl_2$ in an Unknown Mixture

A 0.750-g unknown mixture with calcium chloride gives a 0.455-g precipitate of calcium phosphate. Calculate the percentage of $CaCl_2$ in the unknown mixture.

Solution: In this example, we must relate the mass of precipitate product to the mass of the original $CaCl_2$ reactant.

$$0.455\ \cancel{g\ Ca_3(PO_4)_2} \times \frac{1\ \cancel{mol\ Ca_3(PO_4)_2}}{310.18\ \cancel{g\ Ca_3(PO_4)_2}} \times \frac{3\ \cancel{mol\ CaCl_2}}{1\ \cancel{mol\ Ca_3(PO_4)_2}} \times \frac{110.98\ g\ CaCl_2}{1\ \cancel{mol\ CaCl_2}}$$

$$= 0.488\ g\ CaCl_2$$

Since the sample mixture has a mass of 0.750 g, the percentage of $CaCl_2$ is

$$\frac{\text{mass } CaCl_2}{\text{mass sample}} \times 100\% = \%\ CaCl_2$$

$$\frac{0.488\ \cancel{g}}{0.750\ \cancel{g}} \times 100\% = 65.1\%$$

EQUIPMENT and CHEMICALS

- 250-mL beaker
- 100-mL graduated cylinder
- ring stand & ring
- wire gauze
- clay triangle
- 400-mL beaker
- wash bottle with distilled water
- glass stirring rod & rubber policeman
- calcium chloride, anhydrous $CaCl_2$
- sodium phosphate solution, 0.5 *M* Na_3PO_4
- filter paper
- long-stem funnel (75 mm diameter)
- unknown calcium chloride mixtures (50–70% $CaCl_2$)

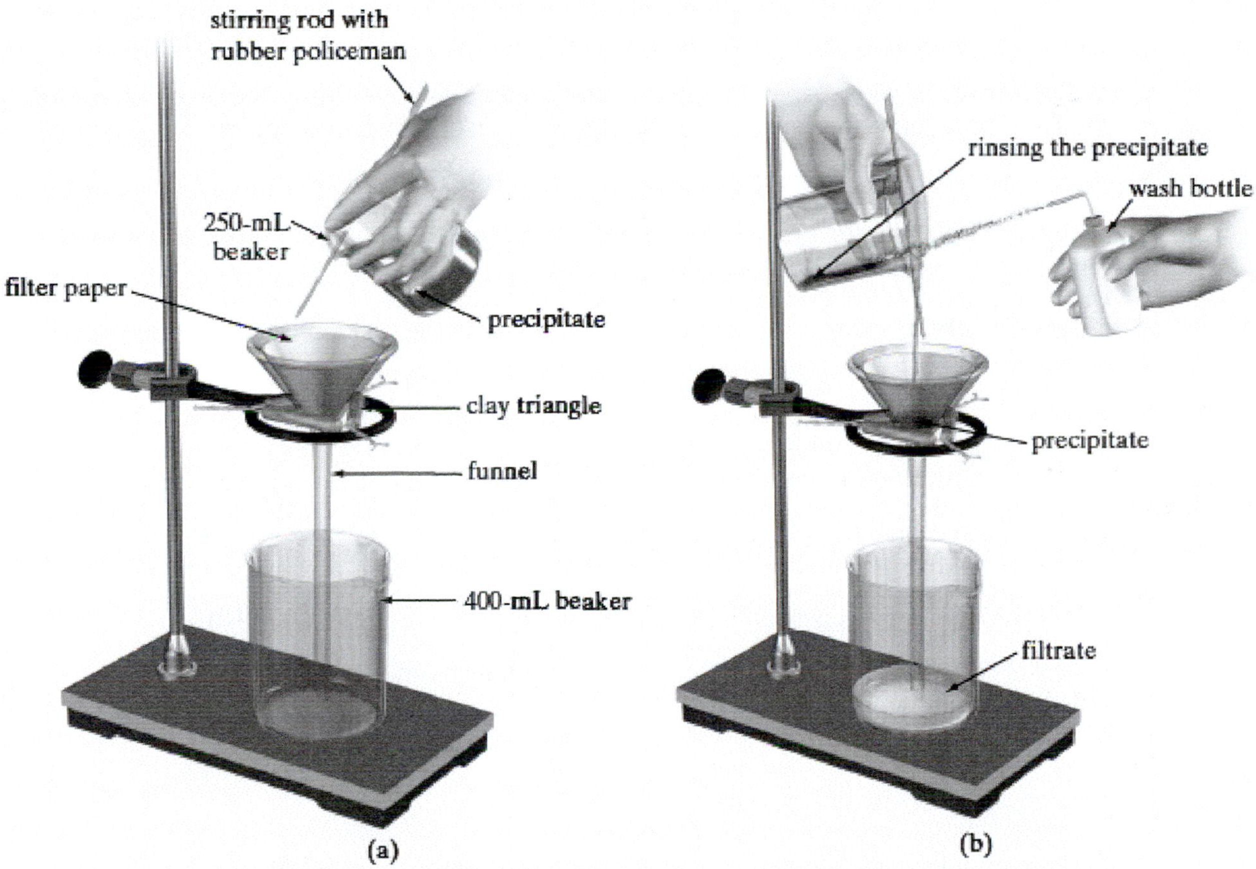

Figure 15.1 Filtration Apparatus (a) When the precipitate settles, pour off the supernate. (b) After the supernate passes through the filter paper, rinse the precipitate into the filter paper using a stream of water from a wash bottle.

PROCEDURE

A. Percent Yield of $Ca_3(PO_4)_2$ from $CaCl_2$

1. Place a 250-mL beaker on the balance, and record the mass. Add ~0.5 g of calcium chloride, and reweigh.

2. Dissolve the $CaCl_2$ sample completely in 50 mL of distilled water. Using a graduated cylinder, add 10 mL of 0.5 *M* Na_3PO_4 solution to the sample in the 250-mL beaker.

3. Support the beaker with a wire gauze on a ring stand. Bring the solution to a gentle boil and turn off the burner. Allow **digestion** of the precipitate until the solution is cool.

   Note: As the precipitate settles from solution, add a few drops of Na_3PO_4 to test for complete precipitation. If the clear solution becomes cloudy, add a few more drops of solution to assure that all the calcium is precipitated.

4. Weigh a disk of filter paper. Prepare a filter paper cone by folding the disk twice. Insert the filter paper into the funnel, and moisten with distilled water using the wash bottle.

5. Assemble a filtering apparatus as shown in Figure 15.1.

6. Without disturbing the precipitate, carefully pour off the supernate into the filter paper, using a stirring rod to guide the flow as shown in Figure 15.1(a). Rinse out the bulk of the precipitate with a stream of water from the wash bottle as shown in Figure 15.1(b). Clean the beaker using a rubber policeman and rinse the residue into the filter paper.

   Note: The precipitate is "jelly-like" and filters quite slowly. To speed filtration, first pour off the clear supernate. Then transfer the precipitate using a minimum amount of wash water. To avoid delay, allow the filtration to continue and begin the unknown part of the experiment.

RECYCLE Chemical Waste

7. After the supernate has passed through the filter, *carefully* remove the paper cone from the funnel. After the precipitate is completely dry, weigh the filter paper and precipitate.

   Note: If an oven is available, dry the precipitate overnight at ~110°C.

8. Calculate the theoretical yield of calcium phosphate from the mass of the calcium chloride. Find the percent yield.

B. Percentage of $CaCl_2$ in an Unknown Mixture

1. Obtain an unknown sample mixture containing calcium chloride, $CaCl_2$. Record the unknown number in the Data Table.

2. Place a 250-mL beaker on the balance and record the mass. Add about 1 g of unknown mixture, and reweigh.

3. Repeat steps 1–7 as in Procedure A; substitute an unknown mixture for calcium chloride.

4. Calculate the mass of calcium chloride in the unknown sample from the mass of precipitate. Find the percentage of calcium chloride in the unknown mixture.

EXPERIMENT 15 NAME ____________________

DATE ____________ SECTION ____________________

*PRELABORATORY ASSIGNMENT**

1. Provide the key term that corresponds to each of the following definitions.

_________ (a) the relationship of quantities (i.e., mass of substance or volume of gas) in a chemical reaction according to the balanced chemical equation

_________ (b) refers to a precipitate containing impurities that are usually soluble

_________ (c) the process of heating a precipitate in aqueous solution to develop larger particles that are easier to filter and free of impurities

_________ (d) the solution above a precipitate after insoluble particles separate from solution

_________ (e) the solution that passes through filter paper in a filtration operation

_________ (f) the amount of product experimentally obtained from a reaction

_________ (g) the amount of product that is calculated from a given amount of reactant

_________ (h) the actual yield compared to the theoretical yield expressed as a percent

Key Terms: actual yield, coprecipitation, digestion, filtrate, percent yield, stoichiometry, supernate, theoretical yield

2. Why must you allow digestion of the precipitate before filtration?

3. What should be done if particles of precipitate appear in the filtrate?

4. What is the purpose of the rubber policeman on the stirring rod?

5. Is it possible to obtain a percent yield of $Ca_3(PO_4)_2$ that is greater than 100%?

6. What are major sources of experimental error?

---

* *Answers in Appendix J*

7. A 0.515-g sample of $CaCl_2$ reacts with aqueous sodium phosphate to give 0.484 g $Ca_3(PO_4)_2$. Refer to Example Exercise 15.1 and show the calculation for the theoretical yield of $Ca_3(PO_4)_2$.

____

What is the percent yield of calcium phosphate, $Ca_3(PO_4)_2$?

____

8. A 0.700-g unknown mixture with $CaCl_2$ reacts with aqueous sodium phosphate to give 0.425 g $Ca_3(PO_4)_2$. Refer to Example Exercise 15.2 and show the calculation for the percentage $CaCl_2$ in the unknown mixture.

____

9. What safety precautions must be observed in this experiment?

EXPERIMENT 15 NAME ______________________

DATE ______________ SECTION ______________________

DATA TABLE

A. Percent Yield of $Ca_3(PO_4)_2$ from $CaCl_2$

mass of beaker + $CaCl_2$ __________ g

mass of beaker __________ g

mass of $CaCl_2$ __________ g

mass of filter paper + $Ca_3(PO_4)_2$ ppt __________ g

mass of filter paper __________ g

mass of $Ca_3(PO_4)_2$ ppt *(actual yield)* __________ g

RECYCLE Chemical Waste

Show the calculation for theoretical yield of $Ca_3(PO_4)_2$ (see Example Exercise 15.1).

mass of $Ca_3(PO_4)_2$ ppt *(theoretical yield)* __________ g

Show the calculation for percent yield of $Ca_3(PO_4)_2$ (see Example Exercise 15.1).

Percent Yield of $Ca_3(PO_4)_2$ __________ %

B. Percentage of $CaCl_2$ in an Unknown Mixture **UNKNOWN #** ______

mass of beaker + unknown mixture	______	g
mass of beaker	______	g
mass of unknown mixture	______	g
mass of filter paper + $Ca_3(PO_4)_2$ ppt	______	g
mass of filter paper	______	g
mass of $Ca_3(PO_4)_2$ ppt	______	g

RECYCLE
Chemical
Waste

Show the calculation for the mass of $CaCl_2$ in the unknown mixture (see Example Exercise 15.2).

mass of $CaCl_2$ ______ g

Show the calculation for the percentage of $CaCl_2$ in the unknown mixture (see Example Exercise 15.2).

Percentage of $CaCl_2$ ______ %

EXPERIMENT 15 NAME ____________________

DATE ____________________ SECTION ____________________

POSTLABORATORY ASSIGNMENT

1. A 0.500-g sample of stannous fluoride gives a 0.590-g precipitate of stannous phosphate. Calculate the theoretical yield and percent yield of $Sn_3(PO_4)_2$.

$$3\ SnF_2(aq) + 2\ Na_3PO_4(aq) \rightarrow Sn_3(PO_4)_2(s) + 6\ NaF(aq)$$

____

2. A 5.000-g sample of toothpaste contains stannous fluoride and gives a 0.075-g precipitate of stannous phosphate. What is the percentage of SnF_2 in the toothpaste sample?

$$3\ SnF_2(aq) + 2\ Na_3PO_4(aq) \rightarrow Sn_3(PO_4)_2(s) + 6\ NaF(aq)$$

____

3. Tooth enamel is the hardest substance in the body and is mainly the mineral hydroxyapatite, $Ca_{10}(PO_4)_6(OH)_2$. If 1.000 g of tooth enamel reacts with hydrochloric acid to give 0.925 g of $CaCl_2$, what is the percentage of hydroxyapatite in tooth enamel?

$$Ca_{10}(PO_4)_6(OH)_2(s) + 20\ HCl(aq) \rightarrow 10\ CaCl_2(aq) + 6\ H_3PO_4(aq) + 2\ H_2O(l)$$

____

5. (optional) Milk of magnesia, $Mg(OH)_2$, is prepared from magnesium sulfate and sodium hydroxide. If a solution containing 100.0 g of $MgSO_4$ is added to a solution with 70.0 g of NaOH, what is the limiting reactant? Calculate the mass of milk of magnesia produced.

$$MgSO_4(aq) + 2\ NaOH(aq) \rightarrow Mg(OH)_2(s) + Na_2SO_4(aq)$$

____

Generating Hydrogen Gas

EXPERIMENT

16

OBJECTIVES

- To determine the experimental molar volume of hydrogen gas at STP.
- To determine the atomic mass and identity for an unknown metal (**X**).
- To gain experience in collecting a gas over water and reading a barometer.

DISCUSSION

The **molar volume** of a gas is the volume occupied by one mole of gas at standard conditions. The theoretical value for the molar volume of an ideal gas at **standard temperature and pressure** (**STP**) is 22.4 liters. A volume of 22.4 L contains Avogadro's number of molecules. The molar volume concept is illustrated in Figure 16.1.

Figure 16.1 The Mole Concept One mole of gas occupies 22.4 L at STP, and contains Avogadro's number of molecules.

In this experiment, magnesium metal reacts with hydrochloric acid according to the equation

$$Mg(s) \quad + \quad 2\,HCl(aq) \quad \rightarrow \quad MgCl_2(aq) \quad + \quad H_2(g)$$

We can find an experimental value for the molar volume of hydrogen gas from the stoichiometry of the reaction. The following example exercise illustrates the calculation of molar volume.

Example Exercise 16.1 • Molar Volume of Hydrogen Gas

A 0.0750 g sample of magnesium metal reacts with hydrochloric acid to produce 77.5 mL of hydrogen gas. The "wet" gas is collected over water at 20 °C and an atmospheric pressure of 763 mm Hg. Calculate the experimental molar volume of hydrogen gas at STP.

Solution: Since the gas is collected over water, both hydrogen gas and water vapor contribute to the total pressure. We find in Table 16.1 the **vapor pressure** of water at 20 °C is 18 mm Hg. The total pressure of the hydrogen gas and water vapor equals the **atmospheric pressure**, that is, 763 mm Hg. Applying **Dalton's law of partial pressures** we have

$$P_{H_2} + P_{H_2O} = P_{atm}$$
$$P_{H_2} + 18 \text{ mm Hg} = 763 \text{ mm Hg}$$
$$P_{H_2} = 763 \text{ mm Hg} - 18 \text{ mm Hg}$$
$$= 745 \text{ mm Hg}$$

Let's prepare a table for the pressure, volume, and temperature data.

Conditions	***P***	***V***	***T***
initial	745 mm Hg	77.5 mL	20 + 273 = 293 K
final	760 mm Hg	V_{STP}	0 + 273 = 273 K

We can correct the volume of H_2 gas to STP using the **combined gas law**.

$$V_{initial} \times P_{factor} \times T_{factor} = V_{STP}$$

The pressure increases, so the volume decreases. The P_{factor} is less than 1. The temperature decreases, so the volume decreases. The T_{factor} is less than 1.

$$77.5 \text{ mL } H_2 \times \frac{745 \text{ \sout{mm Hg}}}{760 \text{ \sout{mm Hg}}} \times \frac{273 \text{ \sout{K}}}{293 \text{ \sout{K}}} = 70.8 \text{ mL } H_2$$

Referring to the above balanced chemical equation, we see that 1 mol Mg metal produces 1 mol H_2 gas; thus,

$$0.0750 \text{ \sout{g Mg}} \times \frac{1 \text{ \sout{mol Mg}}}{24.31 \text{ \sout{g Mg}}} \times \frac{1 \text{ mol } H_2}{1 \text{ \sout{mol Mg}}} = 0.00309 \text{ mol } H_2$$

The molar volume is the ratio of liters of H_2 gas at STP to moles of H_2 gas produced from the reaction. Thus,

$$\frac{70.8 \text{ mL } H_2}{0.00309 \text{ mol } H_2} \times \frac{1 \text{ L}}{1000 \text{ mL}} = 22.9 \text{ L/mol}$$

The experimental molar volume is 22.9 L/mol at STP. This value compares closely with the theoretical molar volume of 22.4 L/mol.

Atomic Mass of an Unknown Metal

After reacting magnesium metal and hydrochloric acid, we will react an unknown metal (**X**) in a similar fashion. The equation for the reaction is

$$\mathbf{X}(s) + 2\,HCl(aq) \rightarrow \mathbf{X}Cl_2(aq) + H_2(g)$$

We can use the concept of molar volume to calculate the atomic mass of the unknown metal as shown in the following example exercise.

Example Exercise 16.2 • Atomic Mass of an Unknown Metal

A 0.215-g sample of unknown metal produces 81.0 mL of hydrogen gas. The "wet" gas is collected over water at 21 °C and an atmospheric pressure of 763 mm Hg. Calculate the atomic mass of the unknown metal (**X**).

Solution: We find in Table 16.1 that the vapor pressure of water at 21 °C is 19 mm Hg. The partial pressure of hydrogen gas is found by applying Dalton's law.

$$P_{H_2} + P_{H_2O} = P_{atm}$$

$$P_{H_2} + 19 \text{ mm Hg} = 763 \text{ mm Hg}$$

$$P_{H_2} = 763 \text{ mm Hg} - 19 \text{ mm Hg}$$

$$= 744 \text{ mm Hg}$$

Let's prepare a table for the pressure, volume, and temperature data.

Conditions	*P*	*V*	*T*
initial	744 mm Hg	81.0 mL	21 + 273 = 294 K
final	760 mm Hg	V_{STP}	0 + 273 = 273 K

We can correct the initial volume of H_2 gas to STP conditions as follows:

$$81.0 \text{ mL } H_2 \times P_{factor} \times T_{factor} = V_{STP}$$

The pressure increases, so the volume decreases. The P_{factor} is less than 1. The temperature decreases, and volume decreases. The T_{factor} is less than 1.

Since the pressure and temperature factors are less than 1, we have

$$81.0\text{ mL } H_2 \quad \times \quad \frac{744\ \cancel{\text{mm Hg}}}{760\ \cancel{\text{mm Hg}}} \quad \times \quad \frac{273\ \cancel{\text{K}}}{294\ \cancel{\text{K}}} \quad = \quad 73.6\text{ mL } H_2$$

According to the balanced equation for the reaction, 1 mole of unknown metal (**X**) produces 1 mole of hydrogen gas. Thus, at STP

$$73.6\ \cancel{\text{mL}}\ H_2 \quad \times \quad \frac{1\ \cancel{\text{L}}}{1000\ \cancel{\text{mL}}} \quad \times \quad \frac{1\ \cancel{\text{mol } H_2}}{22.4\ \cancel{\text{L } H_2}} \quad \times \quad \frac{1\text{ mol } \mathbf{X}}{1\ \cancel{\text{mol } H_2}} \quad = \quad 0.00329\text{ mol } \mathbf{X}$$

The atomic mass of the unknown metal is expressed by the ratio of the mass of sample to the moles of metal.

$$\frac{0.215\text{ g } \mathbf{X}}{0.00329\text{ mol } \mathbf{X}} \quad = \quad 65.3\text{ g/mol}$$

The atomic mass of the unknown metal (**X**) is 65.3 g/mol. If we refer to the periodic table, we identify the unknown metal as zinc (65.39 g/mol).

In this experiment, we will collect hydrogen gas over water. The hydrogen gas displaces water from a graduated cylinder. Figure 16.2 illustrates an apparatus for collecting the gas.

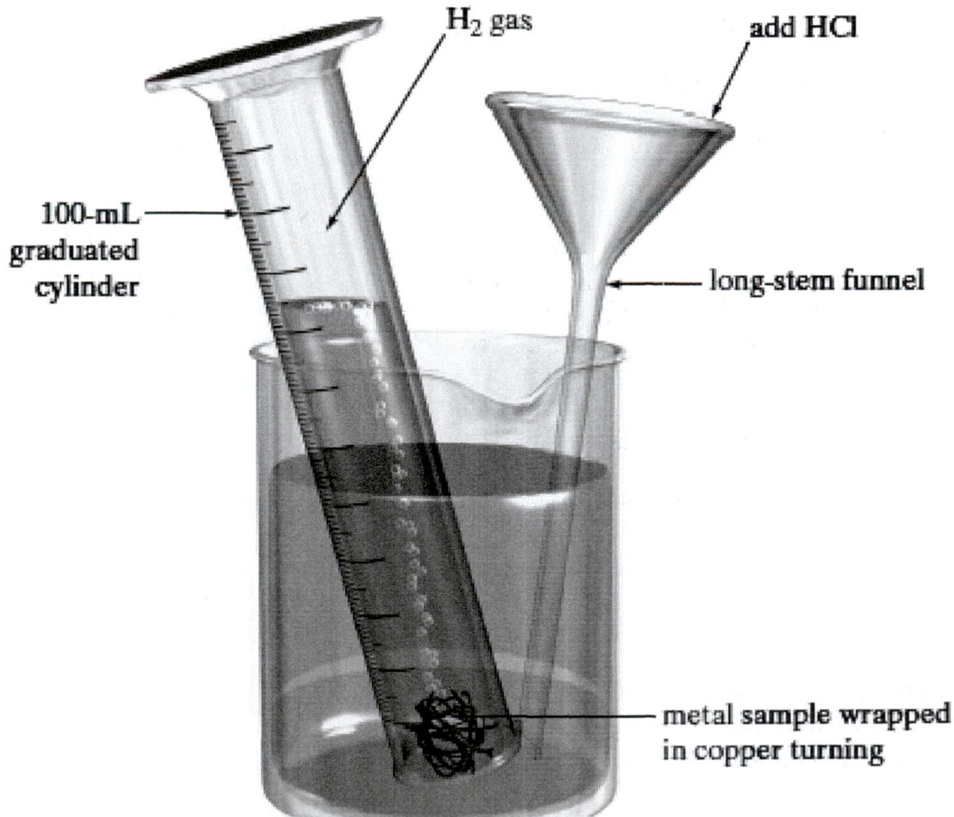

Figure 16.2 Gas Collection Apparatus The volume of hydrogen gas produced equals the volume of water displaced from the graduated cylinder.

EQUIPMENT and CHEMICALS

- 1000-mL beaker
- 100-mL graduated cylinder
- wash bottle with distilled water
- long-stem funnel
- 100-mL beaker
- 110 °C thermometer
- barometer
- milligram balances
- tenth milligram balances (optional)
- magnesium, Mg ribbon
- copper, Cu light turnings
- dilute hydrochloric acid, 6 *M* HCl
- unknown metal samples (**X**)

PROCEDURE

A. Molar Volume of Hydrogen Gas

1. Cut a strip of magnesium ribbon with a 0.070–0.090 g mass (approximately 7–9 cm). Weigh the magnesium metal. Roll the metal ribbon into a compact coil and wrap with strands of copper turnings.

2. Add 700 mL of water into a 1000-mL beaker. Drop the copper-wrapped magnesium metal into the water.

3. Fill a 100-mL graduated cylinder with water. Adjust the water level to the upper rim using water from a wash bottle. Place a small piece of paper towel over the entire rim, and allow it to absorb water. Invert the graduated cylinder over the sink. Carefully put the graduated cylinder into the beaker. As the piece of towel floats free, place the graduated cylinder over the copper-wrapped metal as shown in Figure 16.2.

   Note: It is advisable to practice inverting the graduated cylinder filled with water. If the graduated cylinder loses water upon inversion, the paper towel may be too large. On occasion, the spout in the graduated cylinder is too deep. Exchanging the graduated cylinder solves this problem.

4. Using a long-stem funnel, add 25 mL of dilute hydrochloric acid into the beaker. Gas bubbles are observed when the acid reacts with the magnesium sample. When the reaction is complete, tilt the cylinder vertically and read the bottom of the meniscus. Record the volume (± 0.5 mL) in the Data Table.

   RECYCLE Chemical Waste

5. Place the thermometer into the beaker of water, and observe the temperature. Record the temperature of the hydrogen gas in the Data Table.

   Note: Since the hydrogen gas is collected over water, the temperature of the hydrogen gas is the same as the temperature of the water in the beaker.

6. Read the barometer, and record the atmospheric pressure. Find the vapor pressure of water from Table 16.1.

7. Calculate the molar volume of hydrogen gas at STP.

Table 16.1 Vapor Pressure of Water

Temperature	*Pressure*	*Temperature*	*Pressure*	*Temperature*	*Pressure*
16 °C	14 mm Hg	21 °C	19 mm Hg	26 °C	25 mm Hg
17 °C	15 mm Hg	22 °C	20 mm Hg	27 °C	27 mm Hg
18 °C	16 mm Hg	23 °C	21 mm Hg	28 °C	28 mm Hg
19 °C	17 mm Hg	24 °C	22 mm Hg	29 °C	30 mm Hg
20 °C	18 mm Hg	25 °C	24 mm Hg	30 °C	32 mm Hg

B. Atomic Mass of an Unknown Metal

RECYCLE
Chemical
Waste

1. Obtain a sample of unknown metal (**X**), and record the number in the Data Table.
2. Follow the same steps as in Procedure A.
3. Calculate the atomic mass of the unknown metal (**X**). Refer to the atomic masses in the periodic table, and identify the unknown metal (**X**).

EXPERIMENT 16 NAME ____________________

DATE ____________ SECTION ____________________

*PRELABORATORY ASSIGNMENT**

1. Provide the key term that corresponds to each of the following definitions.

_______ (a) the relationship of quantities (i.e., mass of substance or volume of gas) in a chemical reaction according to the balanced chemical equation

_______ (b) the volume occupied by 1 mole of any gas at STP

_______ (c) a temperature of 273 K and a pressure of 760 mm Hg for a gas

_______ (d) the pressure exerted by air molecules in Earth's atmosphere

_______ (e) the pressure exerted by gaseous vapor above a liquid in a closed container when the rates of evaporation and condensation are equal

_______ (f) the pressure exerted by a mixture of gases is equal to the sum of the pressures exerted by each gas in the mixture

_______ (g) the pressure exerted by a gas is inversely proportional to its volume and directly proportional to its Kelvin temperature

_______ (h) a technique for determining the volume of a gas by measuring the volume of water it displaces

Key Terms: atmospheric pressure, combined gas law, Dalton's law, molar volume, standard temperature and pressure, stoichiometry, vapor pressure, volume by displacement

2. Why does the graduated cylinder remain full of water after inverting?

3. Why must the mass of magnesium be less than 0.09 g in this experiment?

4. How do you tell when the magnesium metal has reacted completely?

5. Explain the meaning of the following terms:

(a) "wet gas"

(b) "dry gas"

---

* *Answers in Appendix J*

6. Hydrogen gas is collected over water at 25 °C and an atmospheric pressure of 766 mm Hg. Refer to Table 16.1 and show the calculation for the partial pressure of hydrogen gas.

7. A 0.0795-g sample of magnesium metal reacts with hydrochloric acid to give 88.5 mL of hydrogen gas at 25 °C and 766 mm Hg. Refer to Example Exercise 16.1 and show the calculation for the volume of hydrogen gas at STP.

   Calculate the moles of hydrogen gas produced.

   Calculate the molar volume of hydrogen gas at STP.

8. What are major sources of experimental error?

9. What safety precautions should be observed in this experiment?

EXPERIMENT 16 NAME ______________________

DATE ______________ SECTION ______________________

DATA TABLE

A. Molar Volume of Hydrogen Gas

mass of magnesium	________ g	________ g
volume of hydrogen gas	________ mL	________ mL
temperature of hydrogen gas	________ °C	________ °C
atmospheric pressure (see barometer)	________ mm Hg	________ mm Hg
vapor pressure of water (see Table 16.1)	________ mm Hg	________ mm Hg
partial pressure of hydrogen gas	________ mm Hg	________ mm Hg

Correct the volume of hydrogen gas to STP for trial 1 (see Example Exercise 16.1).

RECYCLE
Chemical
Waste

volume of hydrogen gas (STP) ________ mL ________ mL

Show the calculation for the moles of hydrogen gas for trial 1 (see Example Exercise 16.1).

moles of hydrogen gas ________ mol ________ mol

Show the calculation for the molar volume of hydrogen gas at STP for trial 1.

Molar volume of hydrogen gas (STP) ________ L/mol ________ L/mol

B. Atomic Mass of an Unknown Metal **UNKNOWN #** ________

mass of metal (**X**)	________ g	________ g
volume of hydrogen gas	________ mL	________ mL
temperature of hydrogen gas	________ °C	________ °C
atmospheric pressure (see barometer)	________ mm Hg	________ mm Hg
vapor pressure of water (see Table 16.1)	________ mm Hg	________ mm Hg
partial pressure of hydrogen gas	________ mm Hg	________ mm Hg

Correct the volume of hydrogen gas to STP for trial 1 (see Example Exercise 16.2).

RECYCLE
Chemical
Waste

volume of hydrogen gas (STP)	________ mL	________ mL

Show the calculation for moles of unknown metal (**X**) for trial 1 (see Example Exercise 16.2).

moles of unknown metal	________ mol	________ mol

Show the calculation for the atomic mass of the unknown metal (**X**) for trial 1.

Atomic mass of the unknown metal (**X**)	________ g/mol	________ g/mol
Identity of the unknown metal (**X**)	________	________

EXPERIMENT 16 NAME ____________________

DATE ____________ SECTION ____________________

POSTLABORATORY ASSIGNMENT

1. A 0.200-g sample of cobalt metal reacted with hydrochloric acid according to the following balanced chemical equation:

$$Co(s) \quad + \quad 2\ HCl(aq) \quad \rightarrow \quad CoCl_2(aq) \quad + \quad H_2(g)$$

The volume of hydrogen gas collected over water was 87.5 mL at 20 °C and a barometer reading of 763 mm Hg. Calculate the STP molar volume for hydrogen.

____________ L/mol

2. A 0.130-g sample of an unknown metal (**X**) reacted with hydrochloric acid according to the following chemical equation:

$$2\ \mathbf{X}(s) \quad + \quad 6\ HCl(aq) \quad \rightarrow \quad 2\ \mathbf{X}Cl_3(aq) \quad + \quad 3\ H_2(g)$$

The volume of hydrogen gas collected over water was 92.0 mL at 20 °C and 763 mm Hg. Calculate the atomic mass of the unknown metal and identify the metal from the periodic table.

____________ g/mol

____________ (**X**)

3. A barometer reads 775 mm Hg. Express the atmospheric pressure in the following units.

   (a) atm

   (b) cm Hg

   (c) in. Hg

4. Applying Boyle's Law Concept:
   As a piston compresses air in a cylinder, the gas pressure (increases/decreases).

5. Applying Charles's Law Concept:
   As a rubber balloon cools in a freezer, the volume (increases/decreases).

6. Applying Gay-Lussac's Law Concept:
   As an automobile tire rolls along the highway, the tire pressure (increases/decreases).

7. (optional) Argon gas has a boiling point of –197 °C. Which of the following diagrams best represents the distribution of argon atoms in a steel sphere at –190 °C?

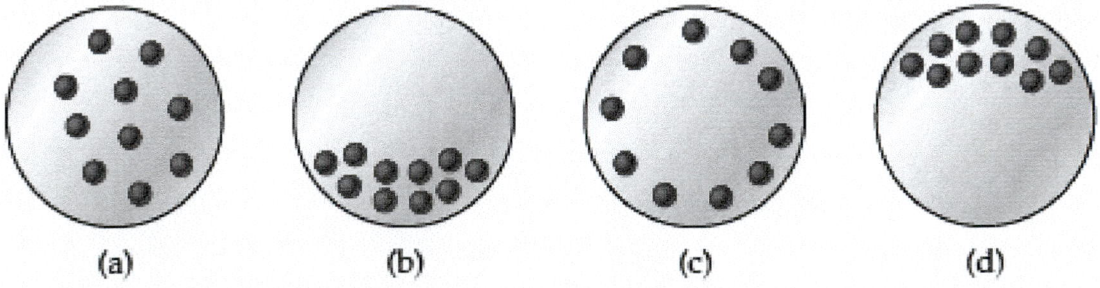

EXPERIMENT

17

Generating Oxygen Gas

OBJECTIVES

- To determine the percentage of potassium chlorate in a known 90.0% mixture.
- To determine the percentage of potassium chlorate in an unknown mixture.
- To gain proficiency in decomposing a compound and collecting a gas over water.

DISCUSSION

When potassium chlorate, $KClO_3$, is heated, it decomposes to potassium chloride and oxygen gas. To assure a safe decomposition, manganese(IV) oxide catalyst, MnO_2, must be mixed with the potassium chlorate. The equation for the decomposition of the black, powdery mixture is

$$2\ KClO_3(s) \xrightarrow[\Delta]{MnO_2} 2\ KCl(s) + 3\ O_2(g)$$

We can determine the volume of oxygen gas produced by measuring the volume of water displaced. This technique is called **volume by displacement**. After correcting the volume of oxygen gas to standard conditions, we can use stoichiometry to relate the volume of oxygen gas to the mass of potassium chlorate in the original sample.

The percentage of potassium chlorate in a sample is found by comparing the mass of $KClO_3$ to the mass of the sample. The following example exercises illustrate the calculation for the percentage potassium chlorate in known and unknown sample mixtures.

Example Exercise 17.1 • % $KClO_3$ in a Known Mixture

A 0.930-g sample of a 90.0% $KClO_3$ mixture is decomposed by heating to give 255 mL of oxygen gas collected over water at 23 °C and an atmospheric pressure of 758 mm Hg. What is the percentage of potassium chlorate in the known sample?

Solution: Since the gas is collected over water, both oxygen gas and water vapor contribute to the total pressure. We find the **vapor pressure** of water from Table 17.1 at 23 °C is 21 mm Hg. The total pressure of the oxygen gas and water vapor equals the **atmospheric pressure**, 758 mm Hg. Applying **Dalton's law of partial pressures**, we have

$$P_{O_2} + P_{H_2O} = P_{atm}$$
$$P_{O_2} + 21 \text{ mm Hg} = 758 \text{ mm Hg}$$
$$P_{O_2} = 758 \text{ mm Hg} - 21 \text{ mm Hg}$$
$$P_{O_2} = 737 \text{ mm Hg}$$

Let's prepare a table for the pressure, volume, and temperature data.

Conditions	*P*	*V*	*T*
initial	737 mm Hg	255 mL	23 + 273 = 296 K
final	760 mm Hg	V_{STP}	0 + 273 = 273 K

We can correct the volume of O_2 gas to STP using the **combined gas law**.

$$V_{initial} \times P_{factor} \times T_{factor} = V_{STP}$$

The pressure increases, so the volume decreases. The P_{factor} is less than 1. The temperature decreases, so the volume decreases. The T_{factor} is less than 1.

$$255 \text{ mL } O_2 \times \frac{737 \cancel{\text{mm Hg}}}{760 \cancel{\text{mm Hg}}} \times \frac{273 \cancel{\text{K}}}{296 \cancel{\text{K}}} = 228 \text{ mL } O_2$$

Referring to the balanced chemical equation, we can see that 2 mol $KClO_3$ (122.55 g/mol) produces 3 mol O_2 gas. If we express **molar volume** as 22,400 mL, we can calculate the mass of $KClO_3$ in the 90.0% mixture.

$$228 \text{ mL } O_2 \times \frac{1 \cancel{\text{mol } O_2}}{22400 \cancel{\text{mL } O_2}} \times \frac{2 \cancel{\text{mol } KClO_3}}{3 \cancel{\text{mol } O_2}} \times \frac{122.55 \text{ g } KClO_3}{1 \cancel{\text{mol } KClO_3}} = 0.832 \text{ g } KClO_3$$

The percentage of $KClO_3$ in the original 0.930 g sample mixture is

$$\frac{\text{mass } KClO_3}{\text{mass sample}} \times 100\% = \% \; KClO_3$$

$$\frac{0.832\ \cancel{g}}{0.930\ \cancel{g}} \times 100\% = 89.5\%$$

Since the result, 89.4%, agrees with the theoretical value, 90.0%, we can conclude that the experimental error is negligible.

Percentage of Potassium Chlorate in an Unknown Mixture

When an unknown mixture containing potassium chlorate is decomposed, the equation for the reaction is the same as above.

Example Exercise 17.2 • % $KClO_3$ in an Unknown Mixture

A 1.145-g sample of an unknown mixture is decomposed by heating. If 205 mL of oxygen gas is collected over water at 22 °C and an atmospheric pressure of 758 mm Hg, what is the percentage of $KClO_3$ in the unknown sample?

Solution: We find in Table 17.1 that the vapor pressure of water at 22 °C is 20 mm Hg. The partial pressure of oxygen gas is found by applying Dalton's law.

$$P_{O_2} + P_{H_2O} = P_{atm}$$
$$P_{O_2} + 20 \text{ mm Hg} = 758 \text{ mm Hg}$$
$$P_{O_2} = 758 \text{ mm Hg} - 20 \text{ mm Hg}$$
$$P_{O_2} = 738 \text{ mm Hg}$$

Let's prepare a table for the pressure, volume, and temperature data.

Conditions	***P***	***V***	***T***
initial	738 mm Hg	205 mL	22 + 273 = 295 K
final	760 mm Hg	V_{STP}	0 + 273 = 273 K

We can correct the initial volume of O_2 gas to STP conditions as follows:

$$V_{initial} \times P_{factor} \times T_{factor} = V_{STP}$$

The pressure increases, so the volume decreases. The P_{factor} is less than 1. The temperature decreases, and volume decreases. The T_{factor} is less than 1.

$$205 \text{ mL } O_2 \times \frac{738\ \cancel{\text{mm Hg}}}{760\ \cancel{\text{mm Hg}}} \times \frac{273\ \cancel{K}}{295\ \cancel{K}} = 184 \text{ mL } O_2$$

According to the balanced equation, we see that 2 mol $KClO_3$ (122.55 g/mol) produces 3 mol O_2 gas. If we express the molar volume as 22,400 mL, we can calculate the mass of $KClO_3$ in the sample as follows:

$$184\ \cancel{mL\ O_2} \times \frac{1\ \cancel{mol\ O_2}}{22400\ \cancel{mL\ O_2}} \times \frac{2\ \cancel{mol\ KClO_3}}{3\ \cancel{mol\ O_2}} \times \frac{122.55\ g\ KClO_3}{1\ \cancel{mol\ KClO_3}} = 0.671\ g\ KClO_3$$

The percentage of $KClO_3$ in the original 1.145 g sample mixture is

$$\frac{\text{mass } KClO_3}{\text{mass sample}} \times 100\% = \%\ KClO_3$$

$$\frac{0.671\ \cancel{g}}{1.145\ \cancel{g}} \times 100\% = 58.6\%$$

Figure 17.1 shows the apparatus for collecting oxygen gas. As the compound decomposes, oxygen gas is produced that displaces water from the Florence flask into a beaker. When the decomposition is complete, no more gas is released and the water level in the beaker remains constant. After a few minutes, the water level actually decreases slightly as the oxygen gas cools in the Florence flask.

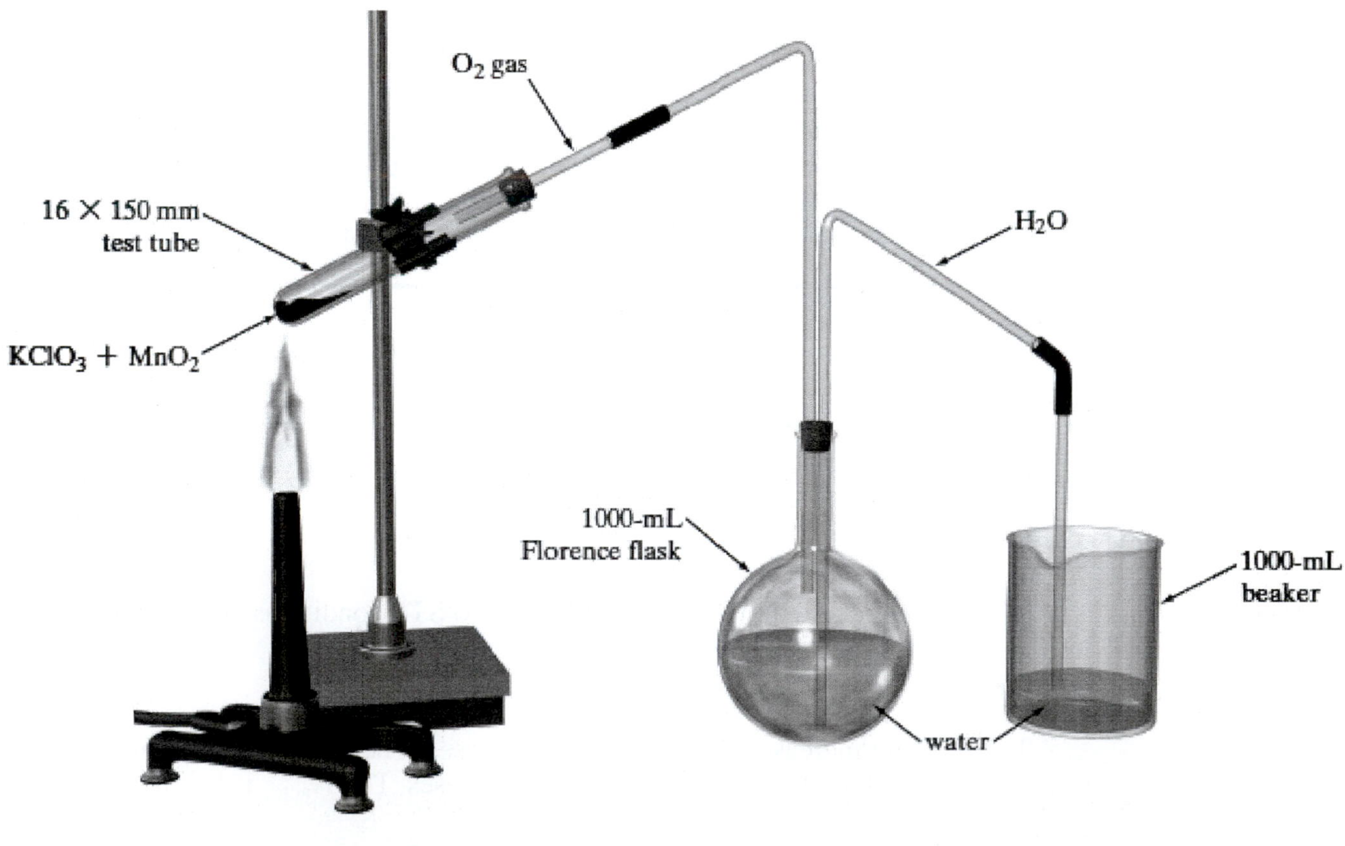

Figure 17.1 Gas Collection Apparatus
When the water level inside the beaker remains constant, the decomposition of $KClO_3$ is complete.

EQUIPMENT and CHEMICALS

- gas collection apparatus (see Figure 17.1)
- 16 x 150 mm test tube
- 1000-mL Florence flask
- 1000-mL beaker
- 110 °C thermometer
- barometer

- gas collection apparatus (see Figure 17.1)
- known potassium chlorate mixture, 90.0% $KClO_3$
- unknown potassium chlorate mixture, 40–80% $KClO_3$

PROCEDURE

A. Percentage of $KClO_3$ in a Known Mixture

1. Weigh a 16 x 150 mm test tube and record the mass. Add about 1 g of the known 90.0% potassium chlorate mixture, and reweigh.

2. Weigh a 1000-mL beaker on a platform balance (see **Appendix B**) and record the mass ± 0.1 g.

3. Set up the apparatus as shown in Figure 17.1. Fill the Florence flask with tap water, and insert the gas collection apparatus. Insert the small rubber stopper into the test tube as shown, and the other end into the 1000-mL beaker.

   Caution: Do not let any of the mixture contact the rubber stopper in the test tube. Ask the Instructor to inspect the apparatus before continuing.

4. Gently heat the test tube and potassium chlorate mixture. Observe water being displaced into the beaker as oxygen gas is produced. As the water level in the beaker increases, heat the test tube more strongly. After the water level remains constant for a couple of minutes, stop heating and allow the test tube to cool for 10 minutes.

RECYCLE Chemical Waste

5. Weigh the 1000-mL beaker containing water on the platform balance. Find the mass of water by difference.

   Note: Since the density of water is 1.00 g/mL, the mass of water in grams is equal to the volume of water in milliliters. Assume the volume of water displaced is equal to the volume of O_2 gas produced from the reaction.

6. Place the thermometer in the water in the Florence flask, and record the temperature of the O_2 gas in the Data Table.

   Note: Since the O_2 gas is collected over water, the temperature of the oxygen gas is the same as the temperature of the water in the flask.

7. Read the barometer, and record the atmospheric pressure. Find the vapor pressure of water from Table 17.1.

8. Correct the volume of oxygen gas to STP. Calculate the mass of potassium chlorate that decomposed and the percentage of $KClO_3$ in the mixture.

Table 17.1 Vapor Pressure of Water

Temperature	*Pressure*	*Temperature*	*Pressure*	*Temperature*	*Pressure*
16 °C	14 mm Hg	21 °C	19 mm Hg	26 °C	25 mm Hg
17 °C	15 mm Hg	22 °C	20 mm Hg	27 °C	27 mm Hg
18 °C	16 mm Hg	23 °C	21 mm Hg	28 °C	28 mm Hg
19 °C	17 mm Hg	24 °C	22 mm Hg	29 °C	30 mm Hg
20 °C	18 mm Hg	25 °C	24 mm Hg	30 °C	32 mm Hg

RECYCLE
Chemical
Waste

B. Percentage of $KClO_3$ in an Unknown Mixture

1. Obtain an unknown sample mixture containing potassium chlorate, $KClO_3$. Record the unknown number in the Data Table.

2. Repeat steps 1–7 as in Procedure A; substitute an unknown mixture for the 90.0% potassium chlorate mixture.

EXPERIMENT 17 NAME ______________________

DATE ______________ SECTION ______________________

*PRELABORATORY ASSIGNMENT**

1. Provide the key term that corresponds to each of the following definitions.

______ (a) the relationship of quantities (i.e., mass of substance or volume of gas) in a chemical reaction according to the balanced chemical equation

______ (b) the volume occupied by 1 mole of any gas at STP

______ (c) a temperature of 273 K and a pressure of 760 mm Hg for a gas

______ (d) the pressure exerted by air molecules in Earth's atmosphere

______ (e) the pressure exerted by gaseous vapor above a liquid in a closed container when the rates of evaporation and condensation are equal

______ (f) the pressure exerted by a mixture of gases is equal to the sum of the pressures exerted by each gas in the mixture

______ (g) the pressure exerted by a gas is inversely proportional to its volume and directly proportional to its Kelvin temperature

______ (h) a technique for determining the volume of a gas by measuring the volume of water it displaces

Key Terms: atmospheric pressure, combined gas law, Dalton's law, molar volume, standard temperature and pressure, stoichiometry, vapor pressure, volume by displacement

2. Why is manganese(IV) oxide added to $KClO_3$ before heating?

3. How do you tell when a $KClO_3$ sample is decomposed completely?

4. If the mass of water produced from heating $KClO_3$ is 345.0 g, what is the volume of water produced? What is the volume of oxygen gas produced?

5. Explain the meaning of the following terms:

(a) "wet gas"

(b) "dry gas"

---

* *Answers in Appendix J*

6. Oxygen gas is collected over water at 23 °C and an atmospheric pressure of 765 mm Hg. Refer to Table 17.1 and show the calculation for the partial pressure of oxygen gas.

7. A 1.000-g sample of the 90.0% $KClO_3$ mixture decomposes with heat to give 275 mL of oxygen gas collected over water at 23 °C and 765 mm Hg. Refer to Example Exercise 17.1 and show the calculation for the volume of oxygen gas at STP.

Calculate the mass of $KClO_3$ in the known mixture.

Calculate the percentage of $KClO_3$ in the known mixture.

8. What are major sources of experimental error?

9. What safety precautions must be observed in this experiment?

EXPERIMENT 17 NAME ____________________

DATE ____________________ SECTION ____________________

DATA TABLE

A. Percentage of $KClO_3$ in a Known Mixture

mass of test tube + $KClO_3$ mixture	________ g	________ g
mass of test tube	________ g	________ g
mass of $KClO_3$ mixture	________ g	________ g
mass of beaker + water (platform balance)	________ g	________ g
mass of beaker (platform balance)	________ g	________ g
mass of water	________ g	________ g
volume of water	________ mL	________ mL
volume of O_2 gas	________ mL	________ mL
temperature of O_2 gas	________ °C	________ °C
atmospheric pressure (see barometer)	________ mm Hg	________ mm Hg
vapor pressure of water (see Table 17.1)	________ mm Hg	________ mm Hg
partial pressure of O_2 gas	________ mm Hg	________ mm Hg

RECYCLE **Chemical Waste**

Show the calculation for the volume of O_2 gas at STP for trial 1 (see Example Exercise 17.1).

volume of O_2 gas (STP) ________ mL ________ mL

Show the calculation for the percentage of $KClO_3$ in the known mixture for trial 1.

Percentage of $KClO_3$ ________ % ________ %

B. Percentage of $KClO_3$ in an Unknown Mixture **UNKNOWN #** ________

mass of test tube + $KClO_3$ mixture	________ g	________ g
mass of test tube	________ g	________ g
mass of $KClO_3$ mixture	________ g	________ g
mass of beaker + water (platform balance)	________ g	________ g
mass of beaker (platform balance)	________ g	________ g
mass of water	________ g	________ g
volume of water	________ mL	________ mL
volume of O_2 gas	________ mL	________ mL
temperature of O_2 gas	________ °C	________ °C
atmospheric pressure (see barometer)	________ mm Hg	________ mm Hg
vapor pressure of water (see Table 17.1)	________ mm Hg	________ mm Hg
partial pressure of O_2 gas	________ mm Hg	________ mm Hg

RECYCLE Chemical Waste

Show the calculation for the volume of O_2 gas at STP for trial 1 (see Example Exercise 17.2).

volume of O_2 gas (STP)	________ mL	________ mL

Show the calculation for the percentage of $KClO_3$ in the unknown mixture for trial 1.

Percentage of $KClO_3$	________ %	________ %
Average percentage of $KClO_3$	________ %	

EXPERIMENT 17 NAME ____________________

DATE ______________ SECTION ____________________

POSTLABORATORY ASSIGNMENT

1. A 1.000-g sample of 90.0% sodium chlorate mixture decomposes with heat to give 315 mL of oxygen gas collected over water at 20 °C and 760 mm Hg. What is the percentage of $NaClO_3$ in the sodium chlorate mixture?

$$2\ NaClO_3(s) \xrightarrow{\Delta} 2\ NaCl(s) + 3\ O_2(g)$$

____

2. A 1.568-g sample of unknown sodium chlorate mixture decomposes with heat to give 272 mL of oxygen gas collected over water at 20 °C and 760 mm Hg. What is the percentage of $NaClO_3$ in the unknown mixture?

$$2\ NaClO_3(s) \xrightarrow{\Delta} 2\ NaCl(s) + 3\ O_2(g)$$

____

3. A barometer reads 752 mm Hg. Express the atmospheric pressure in the following units.

(a) atm

(b) psi

(c) kPa

4. Applying Boyle's Law Concept:
As the air in a cylinder expands, the gas pressure (increases/decreases).

5. Applying Charles's Law Concept:
As a rubber balloon warms on a hot day, the volume (increases/decreases).

6. Applying Gay-Lussac's Law Concept:
As an automobile tire cools after driving, the tire pressure (increases/decreases).

7. (optional) Krypton gas has a boiling point of –152 °C. Which of the following diagrams best represents the distribution of krypton atoms in a steel sphere at –150 °C?

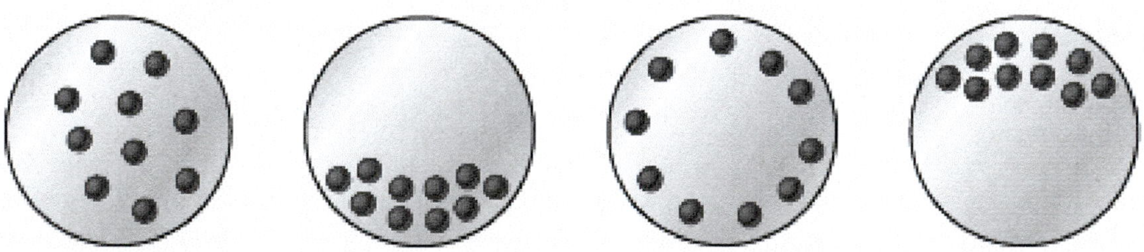

(a) (b) (c) (d)

EXPERIMENT

18

Molecular Models and Chemical Bonds

OBJECTIVES

- To construct models of molecules with single, double, and triple bonds.
- To draw the structural formula for a molecule based on the molecular model.
- To draw the electron dot formula corresponding to the structural formula.
- To draw the structural and electron dot formulas for unknown molecular models.

DISCUSSION

The attraction between two atoms in a molecule is called a chemical bond. In a **covalent bond**, two nonmetal atoms are attracted to each other by sharing valence electrons. The **valence electrons** are the electrons farthest from the nucleus in the outermost portion of an atom. The number of valence electrons can be found by referring to the periodic table. The group number of an element indicates the number of valence electrons. For example, fluorine is in Group VIIA/17 and has seven valence electrons (7 e–).

Example Exercise 18.1 • Valence Electrons and the Periodic Table

Refer to the group number in the periodic table and determine the valence electrons for the following elements: (a) H; (b) C; (c) O; and (d) Cl.

Solution: (a) Hydrogen is in group IA/1; thus, H has **one** valence electron.
(b) Carbon is in Group IVA/14; thus, C has **four** valence electrons.
(c) Oxygen is in Group VIA/16; thus, O has **six** valence electrons.
(d) Chlorine is in Group VIIA/17; thus, Cl has **seven** valence electrons.

In this experiment, we will draw the **structural formula** and **electron dot formula** for molecules after building a model. A model is constructed from spherical balls and connectors, where each ball represents an atom and each connector a single bond. Since a **single bond** shares two electrons, each connector represents an electron pair.

A **double bond** shares two pairs of electrons. A molecular model is constructed using two connectors to represent the double bond. A **triple bond** shares three pairs of electrons. A molecular model is constructed using three connectors to represent the triple bond.

The following example exercises illustrate the structural formula and electron dot formula for molecular models having single, double, and triple bonds.

Example Exercise 18.2 • Structural and Electron Dot Formula for $CHCl_3$

The molecular model of chloroform is sketched below. Draw (a) the structural formula and (b) the electron dot formula. Each atom (excluding H) should be surrounded by an octet of electrons. (c) Verify the electron dot formula by checking the total number of electron dots against the sum of all valence electrons.

chloroform, $CHCl_3$

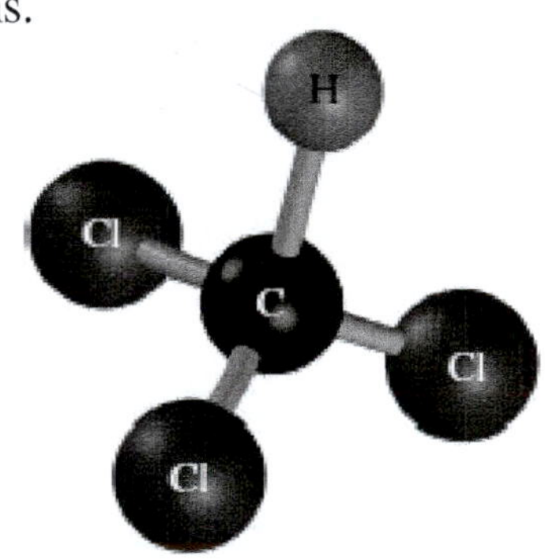

Solution: (a) Each stick represents a single bond, so the structural formula is

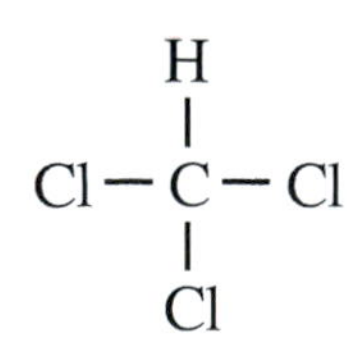

(b) Each dash in the structural formula indicates an electron pair; therefore,

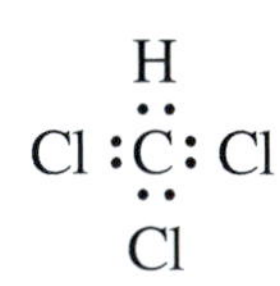

Hydrogen and carbon are complete as shown; two electrons and eight electrons, respectively. However, each chlorine also requires an octet, which we will complete as follows:

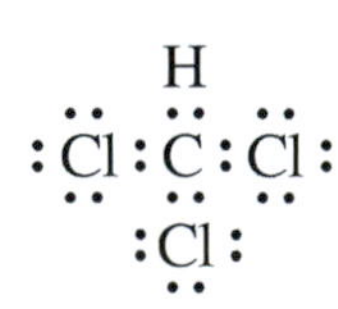

(c) To verify the above electron dot formula, we will find the sum of all valence electrons.

1 H (1 x 1 e–)	=	1 e–
1 C (1 x 4 e–)	=	4 e–
3 Cl (3 x 7 e–)	=	21 e–
sum of valence electrons	=	26 e–

There are 26 valence electrons, and 26 dots were used in the electron dot formula; thus, the formula is verified.

Example Exercise 18.3 • Structural and Electron Dot Formula for H_2CO

A molecular model of formaldehyde is sketched below. Draw the (a) structural formula and (b) electron dot formula. (c) Find the sum of all valence electrons to verify the electron dot formula.

formaldehyde, H_2CO

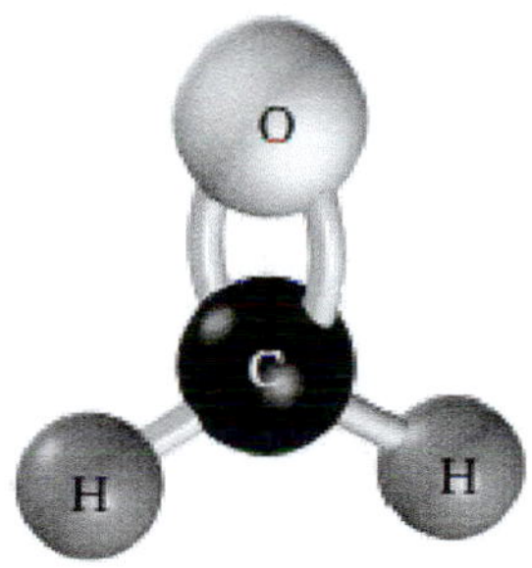

Solution: (a) Two connectors joining the carbon and oxygen atoms represent a double bond. The structural formula can be shown as

```
    O
    ‖
H — C — H
```

(b) Each single bond contains one electron pair, and the double bond two electron pairs.

```
    O
    ::
H : C : H
```

Hydrogen shares two electrons and is stable. Carbon shares a total of eight electrons and satisfies the **octet rule**. Oxygen has only four of the eight electrons necessary to complete the octet. Therefore, we will add two unshared electron pairs.

```
    ..
   :O
    ::
H : C : H
```

(c) We can verify the above electron dot formula as follows:

2 H (2 x 1 e−)	=	2 e−
1 C (1 x 4 e−)	=	4 e−
1 O (1 x 6 e−)	=	6 e−
sum of valence electrons	=	12 e−

The 12 valence electrons equal the 12 electron dots and verify the formula.

Example Exercise 18.4 • Structural and Electron Dot Formula for HCN

A molecular model of hydrogen cyanide is sketched below. Draw (a) the structural formula and (b) the electron dot formula. (c) Verify the electron dot formula.

hydrogen cyanide, HCN

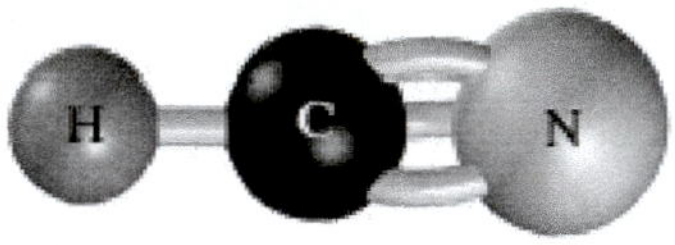

Solution: (a) The three connectors linking the carbon and nitrogen represent a triple pair of electrons.

$$H-C\equiv N$$

(b) We can draw an electron dot formula after realizing the triple bond contains three electron pairs.

H : C ::: N

In the above formula, nitrogen shares only six electrons. Therefore, we must add one unshared electron pair.

H : C ::: N :

(c) Let's verify the preceding electron dot formula.

1 H (1 x 1 e−)	=	1 e−
1 C (1 x 4 e−)	=	4 e−
1 N (1 x 5 e−)	=	5 e−
sum of valence electrons	=	10 e−

The 10 valence electrons verify the 10 e− dots.

EQUIPMENT and CHEMICALS

- Molecular Model Kit
  Student molecular model sets (ISBN: 0-205-08136-3) are available from Prentice Hall @ 1-800-922-0579 (www.prenhall.com).

Directions for Building Molecular Models

When constructing a molecular model, a hole in a ball represents a missing electron that is necessary to satisfy the octet rule. If two balls are joined by a rigid connector, the connector represents a single bond. If two balls are joined by two flexible connectors, the two connectors represent a double bond. If two balls are joined by three flexible connectors, the three connectors represent a triple bond.

one rigid connector	—	single bond (one e– pair)
two flexible connectors	—	double bond (two e– pairs)
three flexible connectors	—	triple bond (three e– pairs)

A molecular model uses different color balls to represent hydrogen, carbon, oxygen, chlorine, bromine, iodine, and nitrogen atoms. The color code for each ball is as follows:

white ball	—	hydrogen atom (one hole)
black ball	—	carbon atom (four holes)
red ball	—	oxygen atom (two holes)
green ball	—	chlorine atom (one hole)
orange ball	—	bromine atom (one hole)
purple ball	—	iodine atom (one hole)
blue ball	—	nitrogen atom (three holes)

Note: For some compounds, it may be difficult to determine the central atom. In this experiment the central atom is shown in **bold** to help build the molecular model; e.g., $H_2\mathbf{O}$.

PROCEDURE

1. Construct molecular models for each compound on the following page. Sketch the model in the Data Table.

2. Draw the structural formula corresponding to the molecular model.

3. Draw the electron dot formula corresponding to the structural formula. Complete the octet by surrounding each atom with 8 electrons (2 electrons for a hydrogen atom).

4. Verify each electron dot formula by summing the valence electrons for the molecule, using the periodic table. This sum should equal the total number of dots in the electron dot formula.

A. Molecular Models with Single Bonds

(a) $\mathbf{H}_2$ (b) $\mathbf{Cl}_2$

(c) $\mathbf{Br}_2$ (d) $\mathbf{I}_2$

(e) H**Cl** (f) H**Br**

(g) **I**Cl (h) **C**H_4

(i) **C**H_2Cl_2 (j) H**O**Cl

(k) $H_2\mathbf{O}_2$ (l) **N**H_3

(m) $\mathbf{N}_2H_4$ (n) **N**H_2**O**H

Note: The central atom is shown in **bold**.

B. Molecular Models with Double Bonds

(a) $\mathbf{O}_2$ (b) $\mathbf{C}_2H_4$

(c) H**O**N**O** (d) H**C**O**O**H

(e) $\mathbf{C}_2H_3Cl$

C. Molecular Models with Triple Bonds

(a) $\mathbf{N}_2$ (b) $\mathbf{C}_2H_2$

(c) H**OC**N

D. Molecular Models with Two Double Bonds

(a) **C**O_2 (b) $\mathbf{C}_3H_4$

E. Unknown Molecular Models

The Instructor will provide models of unknown molecules. Draw the structural formula for each unknown model and the electron dot formula corresponding to each structural formula.

EXPERIMENT 18 NAME ____________________

DATE ______________ SECTION ____________________

*PRELABORATORY ASSIGNMENT**

1. Provide the key term that corresponds to each of the following definitions.

________ (a) the statement that an atom tends to bond in such a way so as to acquire eight electrons in its outer shell

________ (b) the electrons in the outermost *s* and *p* sublevels of an atom that form chemical bonds

________ (c) a bond characterized by the sharing of one or more pairs of valence electrons

________ (d) a bond composed of one electron pair shared between two atoms

________ (e) a bond composed of two electron pairs shared between two atoms

________ (f) a bond composed of three electron pairs shared between two atoms

________ (g) a diagram of a molecule in which each atom is represented by its symbol surrounded by two dots for each pair of bonding or nonbonding electrons

________ (h) a diagram of a molecule that shows the chemical symbol of each atom and a dash representing each pair of bonding electrons

Key Terms: covalent bond, double bond, electron dot formula, octet rule, single bond, structural formula, triple bond, valence electrons

2. Refer to the periodic table and predict the number of valence electrons for atoms of the following elements:

(a) carbon (b) hydrogen

(c) oxygen (d) nitrogen

(e) chlorine (f) bromine

3. What do each of the following represent in the molecular model kit?

(a) one rigid connector (b) two flexible connectors

(c) three flexible connectors (d) white ball

(e) black ball (f) red ball

(g) green ball (h) orange ball

(i) purple ball (j) blue ball

---

* *Answers in Appendix J*

4. Draw the structural formula corresponding to each of the following molecular models.

(a)

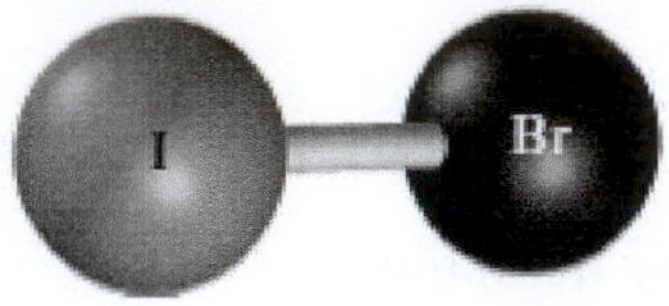

(b)

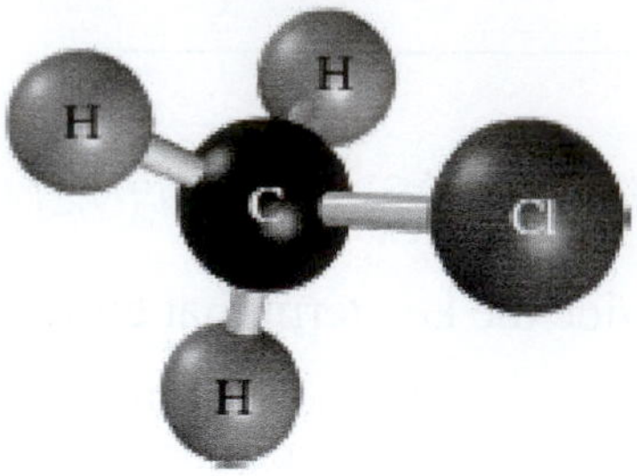

(c)

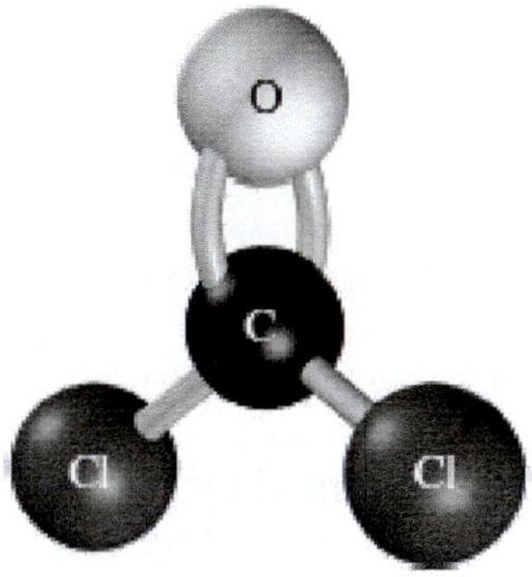

5. Draw the electron dot formula corresponding to the structural formula of the above models.

(a)

(b)

(c)

6. Verify the electron dot formula and sum the valence electrons for the above models.

(a) **IBr** *Total Valence Electrons* ___

(b) $\mathbf{CH_3Cl}$ *Total Valence Electrons* ___

(c) $\mathbf{Cl_2CO}$ *Total Valence Electrons* ___

EXPERIMENT 18 NAME ____________________

DATE ____________________ SECTION ____________________

DATA TABLE

A. Molecular Models with Single Bonds **Model Kit #**_______

Molecule	*Molecular Model*	*Structural Formula*	*Electron Dot Formula*	*Valence Electrons*
(a) **H_2**				
(b) **Cl_2**				
(c) **Br_2**				
(d) **I_2**				

Molecule	*Molecular Model*	*Structural Formula*	*Electron Dot Formula*	*Valence Electrons*
(e) HCl				
(f) HBr				
(g) ICl				
(h) CH_4				
(i) CH_2Cl_2				

Molecule	*Molecular Model*	*Structural Formula*	*Electron Dot Formula*	*Valence Electrons*
(j) HO**Cl** → (j) H**O**Cl				
(k) $H_2\mathbf{O}_2$				
(l) $\mathbf{N}H_3$				
(m) $\mathbf{N}_2H_4$				
(n) $\mathbf{N}H_2OH$				

B. Molecular Models with Double Bonds

Molecule	*Molecular Model*	*Structural Formula*	*Electron Dot Formula*	*Valence Electrons*
(a) $\mathbf{O_2}$				
(b) $\mathbf{C_2}H_4$				
(c) HO**N**O				
(d) H**C**OOH				
(e) $\mathbf{C_2}H_3Cl$				

C. Molecular Models with Triple Bonds

Molecule	*Molecular Model*	*Structural Formula*	*Electron Dot Formula*	*Valence Electrons*
(a) N_2				
(b) C_2H_2				
(c) HOCN				

D. Molecular Models with Two Double Bonds

(a) CO_2				
(b) C_3H_4				

E. Unknown Molecular Models

Molecule	*Molecular Model*	*Structural Formula*	*Electron Dot Formula*	*Valence Electrons*
#1				
#2				
#3				
#4				
#5				

EXPERIMENT 18 NAME ______________________

DATE ______________ SECTION ______________________

POSTLABORATORY ASSIGNMENT

1. Find the number of valence electrons (VE) for the following molecules. Draw the electron dot formula and structural formula for each molecule. The central atom is shown in **bold.**

Molecule	*Electron Dot Formula*	*Structural Formula*
(a) H**O**Br VE =		
(b) **P**I_3 VE =		
(c) **Si**F_4 VE =		
(d) **C**S_2 VE =		
(e) **S**O_2 VE =		

2. Refer to the electron dot formula for each molecule in the preceding question. Using VSEPR theory, predict the electron pair geometry and molecular shape for each molecule.

Molecule	*Electron Pair Geometry*	*Molecular Shape*
(a) **HOBr**		
(b) $\mathbf{PI}_3$		
(c) $\mathbf{SiF}_4$		
(d) $\mathbf{CS}_2$		
(e) $\mathbf{SO}_2$		

3. (optional) A hydrogen ion bonds to an ammonia molecule, $\mathbf{NH}_3$, forming the ammonium ion. Draw the electron dot and structural formulas for $\mathbf{NH}_4^+$. Label the coordinate covalent bond.

Analysis of Saltwater

EXPERIMENT

19

OBJECTIVES

- To observe crystallization of a supersaturated solution.
- To observe the solubility of solid solutes in various solvents.
- To observe the miscibility of water and various solvents.
- To study the factors that affect the rate of dissolving.
- To determine the mass/mass percent concentration and molar concentration of sodium chloride in an unknown saltwater solution.
- To become proficient in pipetting and evaporating a solution to dryness.

DISCUSSION

The ***like dissolves like*** **rule** describes the general principle for dissolving a **solute** in a **solvent**. In general, ionic and polar solutes dissolve in polar solvents, and nonpolar solutes dissolve in nonpolar solvents. Water is a polar solvent and dissolves ionic compounds such as table salt, NaCl, and polar compounds such as table sugar, $C_{12}H_{22}O_{11}$. Hexane is a nonpolar solvent and does not dissolve salt or sugar. Hexane is a good solvent for nonpolar compounds such as grease and oil.

The general principle of *like dissolves like* also dictates whether two liquids are soluble. Liquids that dissolve in one another are said to be **miscible**. If both liquids are polar, they are miscible. Two liquids that are nonpolar also dissolve in one another. However, a polar liquid and a nonpolar liquid are **immiscible**. They are not soluble and separate into two layers.

The amount of solute dissolved in a solution can be expressed in many ways. For example, we can use the terms saturated and unsaturated to indicate the concentration of a solution. A *saturated solution* contains the maximum amount of solute that will dissolve at a given temperature. An *unsaturated solution* contains less than the maximum amount of solute. A **supersaturated solution** is unstable and contains more solute than ordinarily dissolves at a given temperature.

The **mass/mass percent concentration** (symbol **m/m %**) expresses the ratio of the mass of solute to the mass of solution. That is,

$$\frac{\text{mass of solute}}{\text{mass of solution}} \times 100\% = \text{m/m }\%$$

The following Example Exercise 19.1 illustrates a calculation to express the concentration of an aqueous saltwater solution as a mass/mass percent.

Example Exercise 19.1 • Mass/Mass Percent Concentration

A 10.0-mL sample of saltwater solution has a mass of 10.214 g. After evaporating to dryness, the solid NaCl residue weighs 0.305 g. Calculate the mass/mass percent concentration of the solution.

Solution: The mass/mass percent concentration is readily obtained by dividing the mass of NaCl solute by the mass of solution.

$$\frac{0.305 \text{ g NaCl}}{10.214 \text{ g solution}} \times 100\% = 2.99\% \text{ NaCl}$$

The **molar concentration** (symbol ***M***) expresses the number of moles of solute in a liter of solution. Molar concentration is referred to as *molarity*, which we can show as follows.

$$\frac{\text{moles of solute}}{\text{liter of solution}} = M$$

The following Example Exercise 19.2 illustrates a calculation to express the concentration of an aqueous saltwater solution as molarity.

Example Exercise 19.2 • Molar Concentration

Calculate the molar concentration of sodium chloride in the saltwater solution evaporated to dryness in Example Exercise 19.1.

Solution: The molar mass of NaCl is 58.44 g/mol; therefore, the molarity is

$$\frac{0.305 \text{ \sout{g NaCl}}}{10.0 \text{ \sout{mL} solution}} \times \frac{1 \text{ mol NaCl}}{58.44 \text{ \sout{g NaCl}}} \times \frac{1000 \text{ \sout{mL}}}{1 \text{ L}} = \frac{0.522 \text{ mol NaCl}}{1 \text{ L solution}}$$

$$= 0.522\ M \text{ NaCl}$$

EQUIPMENT and CHEMICALS

- 13 x 100 mm test tubes (6) & test tube rack
- 16 x 150 mm test tube
- thin glass stirring rod
- test tube brush
- wash bottle with distilled water
- mortar and pestle
- wire gauze
- evaporating dish
- 250-mL beaker
- 10-mL pipet & bulb
- 100-mL beaker
- ring stand & ring

- sodium acetate trihydrate, solid crystals $NaC_2H_3O_2 \cdot 3\ H_2O$
- potassium permanganate, solid $KMnO_4$
- iodine, solid crystals I_2
- hexane, liquid C_6H_{14}
- methanol, liquid CH_3OH
- acetone, liquid C_3H_6O
- heptane, liquid C_7H_{16}
- ethanol, liquid C_2H_5OH
- rock salt, solid crystals NaCl
- unknown saltwater solutions, 3.00–5.00% NaCl

PROCEDURE

A. Instructor Demonstration – Supersaturation

1. Half-fill a 500-mL Erlenmeyer flask with sodium acetate trihydrate, $NaC_2H_3O_2 \cdot 3\ H_2O$, crystals. Add distilled water to just cover the crystals and place the flask on a hotplate. Heat the flask until the crystals go into solution, swirl, and shut off the heat. Allow the solution to cool undisturbed to room temperature.

    Note: Alternate Directions for Preparing the Supersaturated Solution

    Place 175 g of sodium acetate trihydrate, $NaC_2H_3O_2 \cdot 3\ H_2O$, in a 250-mL Erlenmeyer flask. Add 50 mL distilled water, place the flask on a hotplate, and heat until all the crystals dissolve. Swirl the flask and allow the solution to cool undisturbed to room temperature.

2. Demonstrate supersaturation by dropping a tiny seed crystal of $NaC_2H_3O_2 \cdot 3\ H_2O$ into the solution and have the class observe the results.

    Note: This demonstration is best viewed with the Erlenmeyer flask and solution placed on an illuminated overhead projector. Place a few tiny crystals of sodium acetate trihydrate on the overhead projector for the class to observe. The Instructor may involve the class by asking for a student volunteer to drop a single tiny crystal into the supersaturated solution.

RECYCLE Chemical Waste

B. Solutes and Solvent

1. *Solubility.* Place six *dry* test tubes in a rack. Drop a small crystal of potassium permanganate, $KMnO_4$, into three test tubes as shown in Figure 19.1. Add 10 drops of water to the first, 10 drops of hexane, C_6H_{14}, to the second, and 10 drops of methanol, CH_3OH, to the third. Record whether the crystal is *soluble* or *insoluble*.

   Drop a small crystal of iodine, I_2, into the three remaining test tubes as shown in Figure 19.1. Add 10 drops of water to the first, 10 drops of hexane to the second, and 10 drops of methanol to the third. Shake each test tube and record whether the crystal is *soluble* or *insoluble*.

RECYCLE Chemical Waste

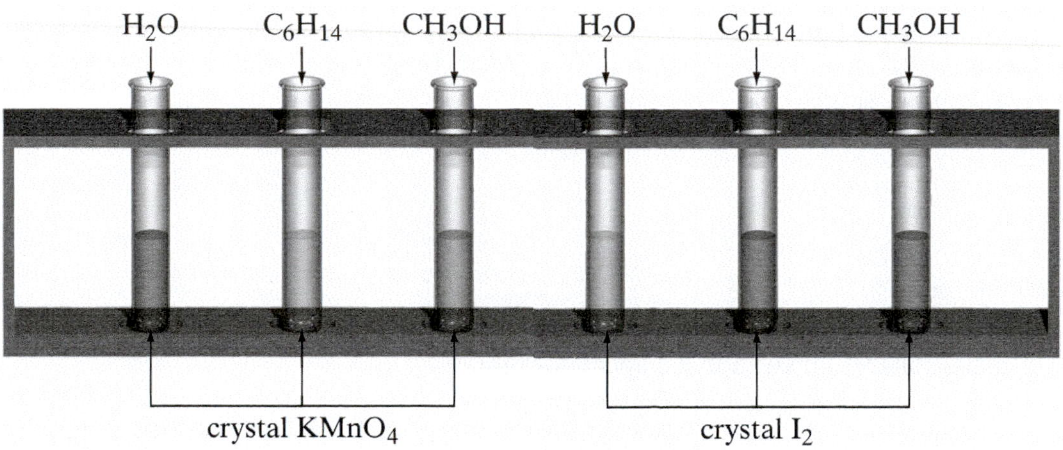

Figure 19.1 Solubility Crystals of $KMnO_4$ and I_2 are soluble in some solvents.

2. *Miscibility.* Put 10 drops of water into each of three test tubes shown in Figure 19.2. Add 10 drops of acetone, C_3H_6O, to the first test tube, 10 drops of heptane, C_7H_{16}, to the second, and 10 drops of ethanol, C_2H_5OH, to the third. Shake each test tube and record whether water and the solvent are *miscible* or *immiscible*.

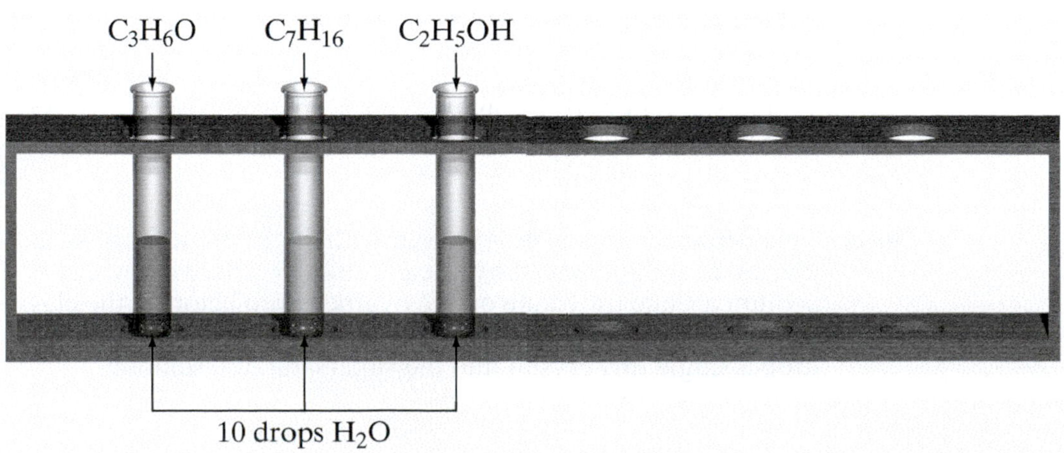

Figure 19.2 Miscibility Water is miscible with some organic solvents.

C. Rate of Dissolving

1. Half-fill a test tube with distilled water. Add a single crystal of rock salt, and later record the length of time required for the crystal to dissolve at ~20 °C.

2. Set up the apparatus shown in Figure 19.3 and add ~150 mL of distilled water to the 250-mL beaker. Half-fill three test tubes with distilled water, and place in the beaker. Heat the water to a boil and shut off the burner.

3. Add a crystal of rock salt to test tube #1 in the beaker. Record the length of time required for the crystal to dissolve at ~100 °C.

4. Add a crystal of rock salt to test tube #2 in the beaker. Stir the solution, and record the length of time required for the crystal to dissolve at ~100 °C.

5. Grind a crystal of rock salt with mortar and pestle. Add the powder to test tube #3. Stir the solution, and record the time required for the powder to dissolve at ~100 °C.

RECYCLE
Chemical
Waste

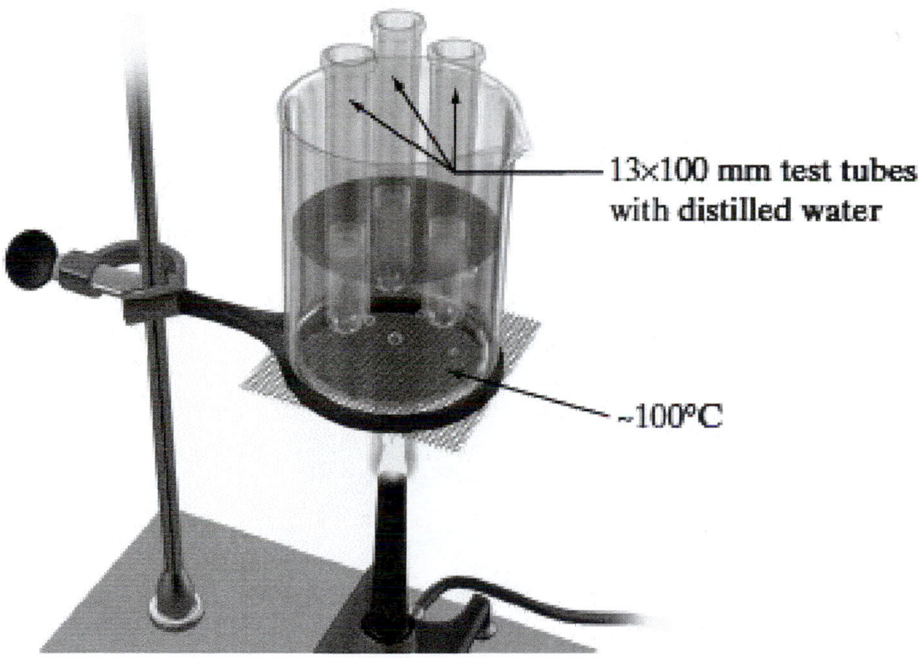

Figure 19.3 Apparatus for Boiling Water The water in the 250-mL beaker heats the distilled water in the test tubes to ~100 °C.

D. Concentration of Sodium Chloride in Saltwater

1. Obtain about 25 mL of unknown saltwater solution in a dry 100-mL beaker, and record the unknown number.

2. Weigh a dry evaporating dish. Condition a pipet with the unknown saltwater solution, and transfer a 10.0-mL sample into the evaporating dish (see **Appendix E**). Reweigh the dish and solution.

3. Add about 200 mL of distilled water into a 250-mL beaker. Place the evaporating dish in the beaker and evaporate the solution to dryness (see Figure 19.5).

4. After evaporation, remove the dish and wipe the bottom of the dish dry. Hold the spout of the dish with crucible tongs over a low flame to dry the last traces of moisture. Allow the dish to cool, and weigh the evaporating dish with the solute residue.

   Note: Do not heat the dish too strongly, as this may cause some of the residue to pop from the dish.

RECYCLE
Chemical Waste

5. Calculate the mass/mass percent and molar concentrations of sodium chloride in the unknown saltwater solution.

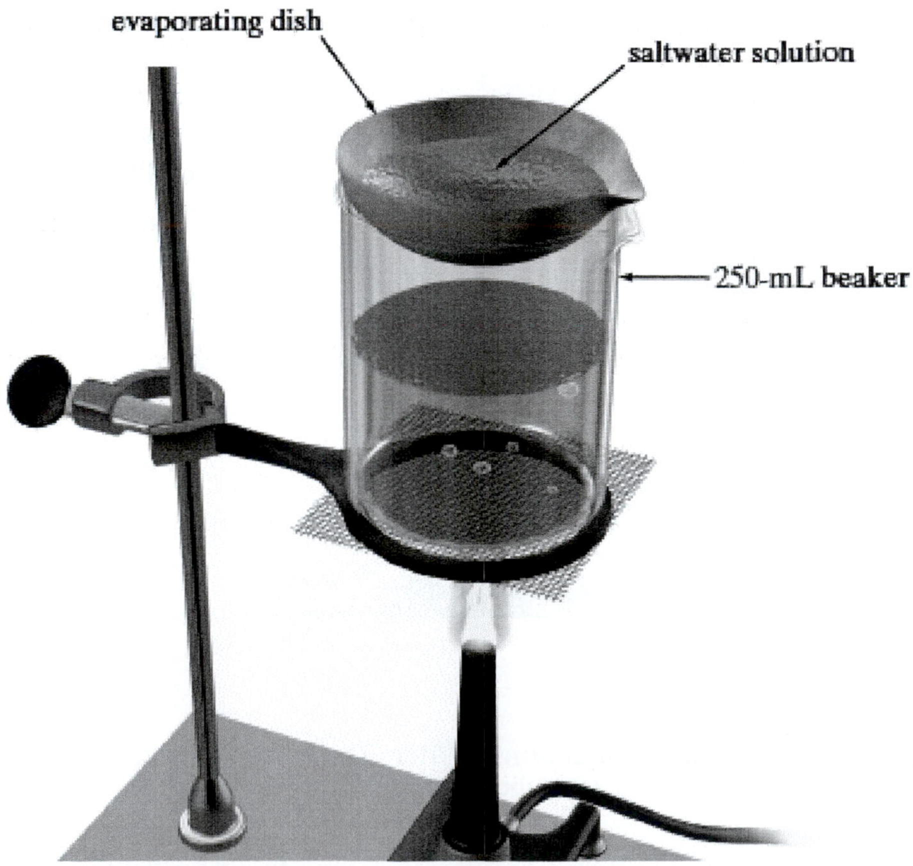

Figure 19.5 Apparatus for Evaporation A saltwater solution is evaporated to dryness using a waterbath. As steam escapes from the beaker, it may be necessary to add more distilled water to the beaker.

EXPERIMENT 19 NAME ____________________

DATE ____________ SECTION ____________________

PRELABORATORY ASSIGNMENT*

1. Provide the key term that corresponds to each of the following definitions.

____________ (a) a homogeneous mixture of a solute dissolved in a solvent

____________ (b) the component of a solution that is the lesser quantity

____________ (c) the component of a solution that is the greater quantity

____________ (d) the general principle that solubility is greatest when the polarity of the solute and solvent are similar

____________ (e) a solution containing more solute than can ordinarily dissolve at a given temperature

____________ (f) an expression that relates the mass of solute dissolved in 100 grams of solution

____________ (g) an expression that relates the moles of solute dissolved in each liter of solution

____________ (h) refers to liquids that dissolve completely in one another

____________ (i) refers to liquids that do not dissolve in one another and separate into layers

____________ (j) to rinse a pipet with a sample liquid to avoid dilution by water on the inside surface

Key Terms: condition, immiscible, *like dissolves like* rule, m/m % concentration, miscible, molar concentration, solute, solution, solvent, supersaturated solution

2. How can you tell if a dark brown iodine crystal is slightly soluble in a solvent?

3. How can you tell if two solvents are miscible? immiscible?

4. Why is *distilled water* used in the waterbath rather than tap water?

* *Answers in Appendix J*

5. A 10.0-mL sample of a saltwater solution containing sodium chloride, NaCl, was evaporated to dryness and gave the following data:

mass of evaporating dish + solution	51.925 g
mass of evaporating dish	41.697 g
mass of evaporating dish + NaCl	42.208 g

(a) Refer to Example Exercise 19.1 and show the calculation for the mass/mass percent concentration of NaCl in the saltwater sample.

(b) Refer to Example Exercise 19.2 and show the calculation for the molar concentration of NaCl (58.44 g/mol) in the saltwater sample.

6. What are major sources of experimental error in analyzing the saltwater solution?

7. What safety precautions must be observed in this experiment?

EXPERIMENT 19 NAME ______________________

DATE ______________ SECTION ______________________

DATA TABLE

A. Instructor Demonstration – Supersaturation

<u>*Observation*</u>

B. Solutes and Solvents

1. *Solubility*

SOLVENT

SOLUTE	*water* H_2O	*hexane* C_6H_{14}	*methanol* CH_3OH
$KMnO_4$			
I_2			

RECYCLE Chemical Waste

2. *Miscibility*

	acetone C_3H_6O	*heptane* C_7H_{16}	*ethanol* C_2H_5OH
H_2O			

C. Rate of Dissolving

Temperature	*Stirring*	*Particle Size*	*Time*
~20 °C	No	Crystal	
~100 °C	No	Crystal	
~100 °C	Yes	Crystal	
~100 °C	Yes	Powder	

D. Concentration of Sodium Chloride in Saltwater **UNKNOWN #** ______

volume of saltwater solution	______ mL	______ mL
mass of evaporating dish + solution *(before heating)*	______ g	______ g
mass of evaporating dish	______ g	______ g
mass of saltwater solution	______ g	______ g
mass of evaporating dish + NaCl solute *(after heating)*	______ g	______ g
mass of NaCl solute	______ g	______ g

RECYCLE
Chemical
Waste

Show the calculation for percent concentration of NaCl for trial 1 (see Example Exercise 19.1).

Mass/mass percent concentration of NaCl	______ %	______ %

Show the calculation for molar concentration of NaCl for trial 1 (see Example Exercise 19.2).

Molarity of NaCl	______ *M*	______ *M*
Average molarity of NaCl	______ *M*	

EXPERIMENT 19 NAME ______________________

DATE ______________ SECTION ______________________

POSTLABORATORY ASSIGNMENT

1. Indicate whether the solute solid is *soluble* or *insoluble* in each solvent liquid.

SOLUTE	SOLVENT: polar liquid, H_2O	SOLVENT: nonpolar liquid, C_6H_{14}
salt, $NaCl$		
sugar, $C_6H_{12}O_6$		
cholesterol, $C_{27}H_{46}O$		

2. Based on the "like dissolves like" rule, predict whether the following vitamins are *soluble* or *insoluble* in water.

   (a) vitamin A, $C_{20}H_{30}O$ ______________

   (b) vitamin B-2, $C_{17}H_{20}N_4O_6$ ______________

   (c) vitamin C, $C_6H_8O_6$ ______________

3. Indicate whether the solute liquid is *miscible* or *immiscible* in each solvent liquid.

SOLUTE	SOLVENT: polar liquid, H_2O	SOLVENT: nonpolar liquid, C_6H_{14}
methanol, CH_3OH		
pentane, C_5H_{12}		

4. Based on the "like dissolves like" rule, predict whether the following solvents are *miscible* or *immiscible* with water.

   (a) gasoline, C_8H_{18} ______________

   (b) rubbing alcohol, C_3H_7OH ______________

   (c) chloroform, $CHCl_3$ ______________

5. Indicate whether the rate of dissolving *increases* or *decreases* for each of the following.

(a) heat the solution ______________

(b) stir the solution ______________

(c) grind the solute ______________

6. A 50.0-mL sample of sodium carbonate solution was found to have a mass of 50.225 g. When the solution was heated to dryness, the mass of the Na_2CO_3 residue was 2.799 g. Calculate the (a) mass/mass percent concentration, and (b) molar concentration of the solution.

(a) ______________

(b) ______________

7. (optional) A normal saline solution for injecting hospital patients is 0.90% NaCl.

(a) What is the mass of NaCl dissolved in 50.0 g of 0.90% saline solution?

____

(b) What is the molar concentration of normal 0.90% saline (58.44 g/mol) solution? (Assume the density of the normal saline solution is 1.01 g/mL.)

____

EXPERIMENT

20

Analysis of Vinegar

OBJECTIVES

- To prepare a standard sodium hydroxide solution.
- To determine the molar concentration and mass/mass percent concentration of acetic acid in an unknown vinegar solution.
- To gain proficiency in the laboratory technique of titration.

DISCUSSION

In this experiment, we will neutralize an acidic solution of vinegar using a basic solution of sodium hydroxide. We determine the amount of sodium hydroxide necessary by performing a **titration** using a buret. When the acid is completely neutralized by the base, the titration stops. This is called the **endpoint** in the titration and is signaled when an **indicator** changes color. At the endpoint in the titration, a single drop of base is sufficient to bring about a permanent color change. Figure 20.1 illustrates a typical titration.

Preparation of a Standard Sodium Hydroxide Solution

We begin by diluting 6 *M* NaOH with water. Since diluting NaOH provides only an approximate concentration and it is necessary to know the concentration of NaOH precisely, we will prepare a **standard solution** by titration.

First, we weigh crystals of potassium hydrogen phthalate, $KHC_8H_4O_4$ (abbreviated KHP). After dissolving the KHP crystals in water, we will titrate the acid solution with NaOH according to the following equation.

$$\text{KHP(aq)} + \text{NaOH(aq)} \rightarrow \text{KNaP(aq)} + \text{H}_2\text{O(l)}$$

Example Exercise 20.1 • Molar Concentration of Standard NaOH

A 0.515 g sample of KHP (204.23 g/mol) is dissolved in water and requires 12.75 mL of NaOH solution to reach a faint pink endpoint. Find the molarity of the NaOH solution.

Solution: Referring to the preceding equation for the reaction and applying the rules of stoichiometry, we have

$$0.515\ \cancel{\text{g KHP}} \times \frac{1\ \cancel{\text{mol KHP}}}{204.23\ \cancel{\text{g KHP}}} \times \frac{1\ \text{mol NaOH}}{1\ \cancel{\text{mol KHP}}} = 0.00252\ \text{mol NaOH}$$

The molarity of the NaOH is found as follows:

$$\frac{0.00252\ \text{mol NaOH}}{12.75\ \cancel{\text{mL}}\ \text{solution}} \times \frac{1000\ \cancel{\text{mL}}}{1\ \text{L}} = \frac{0.198\ \text{mol NaOH}}{1\ \text{L solution}} = 0.198\ M\ \text{NaOH}$$

In this example, the concentration of the standard NaOH solution is 0.198 *M*.

Titration of Acetic Acid in a Vinegar Unknown

After preparing a standard NaOH solution, we will determine the concentration of acetic acid in an unknown vinegar solution. A sample of vinegar will be titrated with NaOH to a permanent endpoint. The equation for the reaction is

$$HC_2H_3O_2(aq) \ + \ NaOH(aq) \ \rightarrow \ NaC_2H_3O_2(aq) \ + \ H_2O(l)$$

The following example exercise illustrates the calculation for the percentage of acetic acid in an unknown vinegar sample.

Example Exercise 20.2 • Percentage of Acetic Acid in Vinegar

The titration of a 10.0-mL vinegar sample requires 38.05 mL of standard 0.198 *M* NaOH. Calculate the (a) molarity and (b) mass/mass percent concentration of acetic acid.

Solution: We can calculate the moles of acetic acid from the moles of NaOH solution:

$$38.05 \ \cancel{\text{mL solution}} \times \frac{0.198 \ \cancel{\text{mol NaOH}}}{1000 \ \cancel{\text{mL solution}}} \times \frac{1 \text{ mol } HC_2H_3O_2}{1 \ \cancel{\text{mol NaOH}}} = 0.00753 \text{ mol } HC_2H_3O_2$$

(a) The molar concentration of $HC_2H_3O_2$ is

$$\frac{0.00753 \text{ mol } HC_2H_3O_2}{10.0 \ \cancel{\text{mL}} \text{ solution}} \times \frac{1000 \ \cancel{\text{mL}}}{1 \text{ L}} = \frac{0.753 \text{ mol } HC_2H_3O_2}{1 \text{ L solution}}$$

$$= 0.753 \ M \ HC_2H_3O_2$$

(b) To calculate the m/m % concentration, we must know the density of the vinegar (1.01 g/mL) and the molar mass of acetic acid (60.06 g/mol).

$$\frac{0.753 \ \cancel{\text{mol } HC_2H_3O_2}}{1000 \ \cancel{\text{mL solution}}} \times \frac{60.06 \text{ g } HC_2H_3O_2}{1 \ \cancel{\text{mol } HC_2H_3O_2}} \times \frac{1 \ \cancel{\text{mL solution}}}{1.01 \text{ g solution}} \times 100\%$$

$$= 4.48\% \ HC_2H_3O_2$$

EQUIPMENT and CHEMICALS

- buret stand & clamp
- ring stand & utility clamp (optional)
- 50-mL buret
- small, plastic funnel (optional)
- 150-mL beaker
- graduated cylinder
- 1000-mL Florence flask w/stopper
- 125-mL Erlenmeyer flasks (3)
- 10-mL pipet & bulb
- 100-mL beaker
- wash bottle with distilled water
- dilute sodium hydroxide, 6 *M* NaOH
- potassium hydrogen phthalate, solid $KHC_8H_4O_4$ (KHP)
- phenolphthalein indicator (or, cresol red indicator)
- unknown vinegar solution, 3.00–5.00% $HC_2H_3O_2$

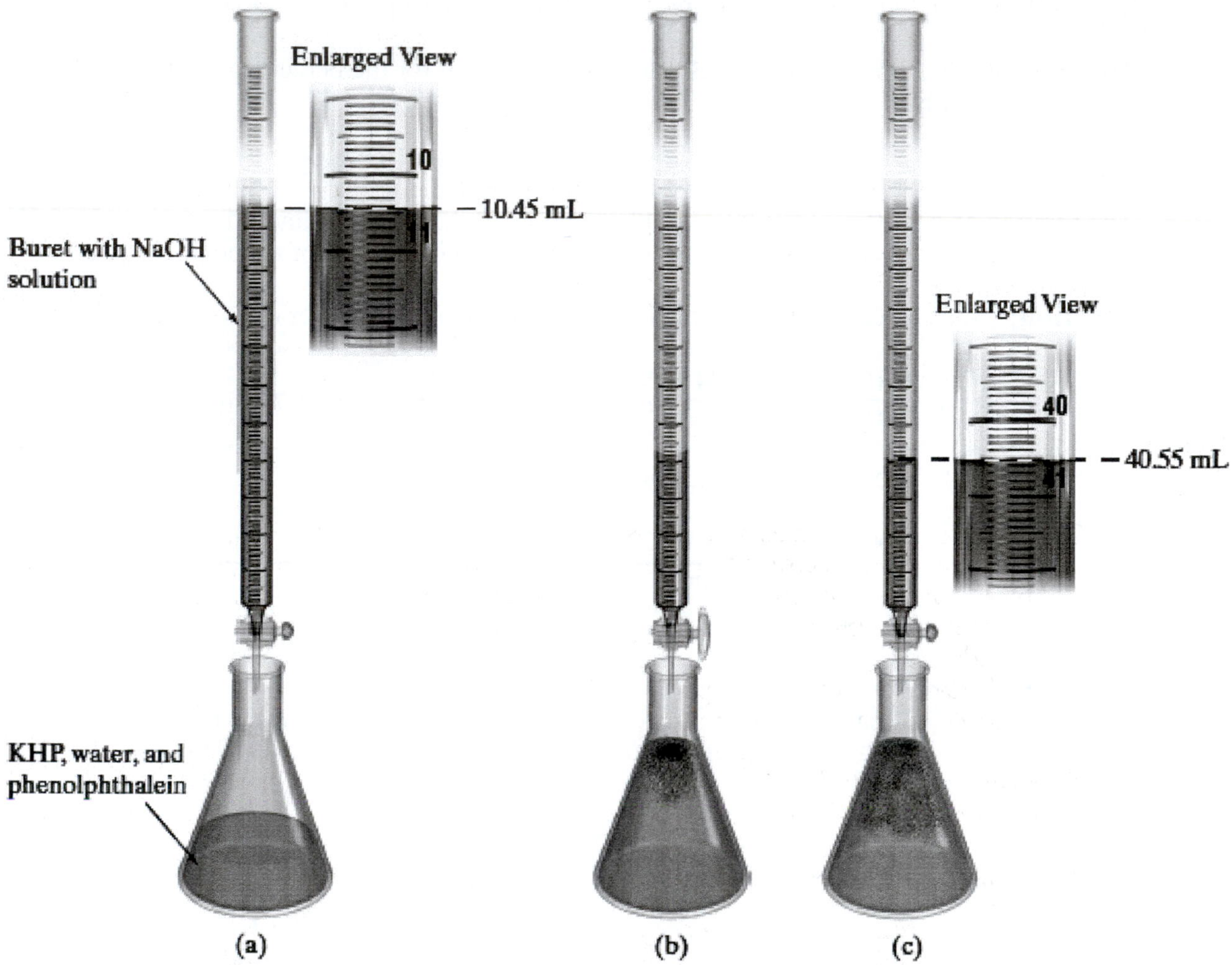

Figure 20.1 Apparatus for Titrating an Acid with a Base
(a) Read the initial volume of NaOH in the buret (10.45 mL). (b) A flash of color indicates nearing the endpoint. (c) A permanent color signals the endpoint for the titration (40.55 mL). In the example shown, the volume of NaOH is 40.55 mL – 10.45 mL = 30.10 mL.

PROCEDURE

A. Preparation of Sodium Hydroxide Solution

1. Half-fill a 1000-mL Florence flask with ~500 mL of distilled water. Measure ~15 mL of 6 *M* NaOH into a graduated cylinder and pour the NaOH into the Florence flask. Stopper the flask, and carefully swirl to mix the solution.

2. Condition a buret with NaOH solution from the Florence flask. Use a small funnel and half-fill the buret with NaOH. Allow some solution to pass through the buret tip into a 150-mL beaker, and empty the remainder into the sink.

3. Close the stopcock, and fill the buret with NaOH solution from the Florence flask.

   Note: Carefully add NaOH solution to the funnel so as to not overfill the buret.

4. Label the 125-mL Erlenmeyer flasks #1, #2, and #3. Precisely weigh ~0.5 g of KHP into each of the flasks. Add ~25 mL of distilled water to each flask, and heat as necessary to dissolve the KHP crystals.

   Note: If a digital electronic balance is available, the Instructor may direct students to tare weigh the KHP samples.

5. Titrate three KHP samples as follows:
   - Drain NaOH through the tip of the buret to clear any air bubbles.
   - Position Erlenmeyer flask #1 under the buret as shown in Figure 20.1.
   - Record the initial buret reading (± 0.05 mL).
   - Add a drop of indicator to the flask.
   - Titrate with NaOH to a permanent endpoint while slowly swirling the flask.
   - Record the final buret reading (± 0.05 mL).

6. Refill the buret with NaOH solution, record the initial buret reading, add a drop of indicator to flask #2, titrate the KHP sample, and record the final buret reading.

7. Refill the buret with NaOH solution, record the initial buret reading, add a drop of indicator to flask #3, titrate the KHP sample, and record the final buret reading.

RECYCLE
Chemical Waste

8. Calculate the molarity of the NaOH solution for each trial. Record the average molarity of NaOH in the Data Table of Procedure B.

Note: ***Save the NaOH in the Florence flask for Procedure B.***

B. Titration of Acetic Acid in Vinegar

1. Obtain ~50 mL of vinegar solution in a dry 100-mL beaker. Record the unknown number in the Data Table.

2. Condition a pipet with unknown vinegar solution, and transfer a 10.0-mL sample into each 125-mL flask (see **Appendix E**). Add ~25 mL of distilled water into each flask.

   Note: It is not necessary to use dry flasks.

3. Titrate three vinegar samples as follows:
   - Position Erlenmeyer flask #1 under the buret.
   - Record the initial buret reading (± 0.05 mL).
   - Add a drop of indicator to the flask.
   - Titrate with NaOH to a permanent endpoint while slowly swirling the flask.
   - Record the final buret reading (± 0.05 mL).

RECYCLE
Chemical
Waste

4. Refill the buret with NaOH solution, record the initial buret reading, add a drop of indicator to flask #2, titrate the vinegar sample, and record the final buret reading.

5. Refill the buret with NaOH solution, record the initial buret reading, add a drop of indicator to flask #3, titrate the vinegar sample, and record the final buret reading.

6. Calculate the molarity of acetic acid, $HC_2H_3O_2$, in the unknown vinegar solution.

7. Convert the molarity of $HC_2H_3O_2$ (60.06 g/mol) to mass/mass percent concentration. Assume the density is 1.01 g/mL for the unknown vinegar solution.

Note: ***When the titrations are complete, rinse the buret and glassware with distilled water to remove all traces of NaOH solution.***

EXPERIMENT 20 NAME ____________________

DATE ____________ SECTION ____________________

*PRELABORATORY ASSIGNMENT**

1. Provide the key term that corresponds to each of the following definitions.

_________ (a) a procedure for obtaining the mass of a sample directly by first placing a container on an electronic balance and setting the balance to zero; second, add a sample to the container and record the mass of sample directly

_________ (b) a procedure for delivering a measured volume of solution using a buret

_________ (c) the clear lens at the surface of a liquid inside a buret

_________ (d) to rinse a pipet or buret with a sample liquid to avoid dilution by water on the inside surface

_________ (e) a substance that undergoes a color change according to the pH of a solution

_________ (f) the stage in a titration when the indicator changes color

_________ (g) a solution whose concentration has been established precisely

_________ (h) an expression that relates the moles of solute dissolved in each liter of solution

_________ (i) an expression that relates the mass of solute dissolved in 100 grams of solution

Key Terms: condition, endpoint, indicator, m/m % concentration, meniscus, molar concentration, standard solution, tare weighing, titration

2. Observe and record the following buret readings.

(a)

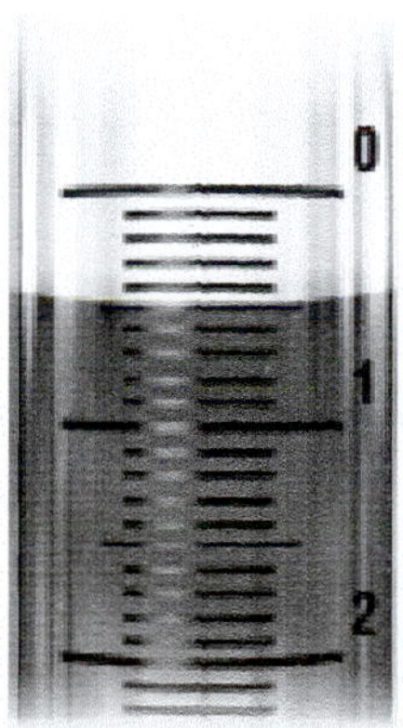

(b)

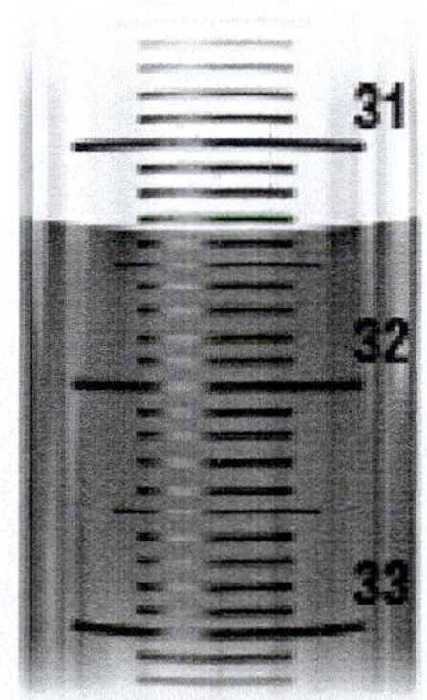

3. How can you tell that you are nearing the endpoint in a titration?

4. What volume of NaOH is required to permanently change the indicator at the endpoint?

---

* *Answers in Appendix J*

5. If KHP sample #1 requires 19.90 mL of NaOH solution to reach an endpoint, what volume should be required for samples #2 and #3?

6. If vinegar sample #1 requires 29.05 mL of NaOH solution to reach an endpoint, what volume should be required for samples #2 and #3?

7. A 0.875-g sample of KHP (204.23 g/mol) is dissolved in water and titrated with 20.75 mL of NaOH solution to a permanent endpoint. Refer to Example Exercise 20.1 and calculate the molarity of the NaOH solution.

8. A 10.0-mL vinegar sample is pipetted into an Erlenmeyer flask and titrated with 28.85 mL of 0.206 *M* NaOH to a permanent endpoint.

   (a) Refer to Example Exercise 20.2 and calculate the molarity of the acetic acid in the vinegar.

   (b) Assume the density of the vinegar solution is 1.01 g/mL and find the mass/mass percent concentration of acetic acid, $HC_2H_3O_2$ (60.06 g/mol), in the unknown vinegar sample.

9. Which of the following is a serious source of experimental error?
   (a) The Erlenmeyer flasks are not dry before weighing the KHP samples.
   (b) The KHP samples are dissolved in 50 mL (*not 25 mL*) of distilled water.
   (c) The sodium hydroxide is not mixed completely in the Florence flask.
   (d) The buret is not conditioned.
   (e) Bubbles are not cleared from the tip of the buret.
   (f) Two drops (*not one drop*) of indicator is used.
   (g) Disposing of the standard NaOH solution before titrating the vinegar samples.
   (h) The Erlenmeyer flasks are not dry before pipetting the vinegar samples.

10. What safety precautions should be observed in this experiment?

EXPERIMENT 20 NAME ____________________

DATE ____________ SECTION ____________________

DATA TABLE

A. Preparation of Sodium Hydroxide Solution

mass of Erlenmeyer flask + KHP	______ g	______ g	______ g
mass of Erlenmeyer flask	______ g	______ g	______ g
mass of KHP	______ g	______ g	______ g
final buret reading	______ mL	______ mL	______ mL
initial buret reading	______ mL	______ mL	______ mL
volume of NaOH	______ mL	______ mL	______ mL

Show the calculation of molarity of NaOH for trial 1 (see Example Exercise 20.1).

RECYCLE
Chemical
Waste

Molarity of NaOH	______ *M*	______ *M*	______ *M*
Average molarity of NaOH		______ *M*	

B. Titration of Acetic Acid in Vinegar **UNKNOWN #** ______

Average molarity of NaOH (see Procedure A)		______ *M*	
volume of vinegar	______ mL	______ mL	______ mL
final buret reading	______ mL	______ mL	______ mL
initial buret reading	______ mL	______ mL	______ mL
volume of NaOH	______ mL	______ mL	______ mL

Show the calculation for the molarity of acetic acid for trial 1 (see Example Exercise 20.2).

RECYCLE
Chemical
Waste

Molarity of $HC_2H_3O_2$	______ *M*	______ *M*	______ *M*
Average molarity of $HC_2H_3O_2$		______ *M*	

Show the calculation for the percent concentration of acetic acid for trial 1. (Assume the density of the vinegar solution is 1.01 g/mL.)

Mass/mass percent $HC_2H_3O_2$	______ %	______ %	______ %
Average mass/mass percent $HC_2H_3O_2$		______ %	

EXPERIMENT 20 NAME ______________________

DATE ______________ SECTION ______________________

POSTLABORATORY ASSIGNMENT

1. A standard nitric acid solution is prepared using 0.425 g of sodium carbonate, Na_2CO_3. Find the molarity of the acid if 33.25 mL are required to reach a permanent endpoint.

$$2\ HNO_3\ (aq) + Na_2CO_3\ (s) \rightarrow 2\ NaNO_3\ (aq) + H_2O(l) + CO_2\ (g)$$

____

2. A 10.0-mL sample of household ammonia solution required 27.50 mL of 0.241 *M* HNO_3 for neutralization. Calculate (a) the molar concentration of the ammonia and (b) the mass/mass percent concentration of ammonia (17.04 g/mol), given a solution density of 0.985 g/mL.

$$HNO_3\ (aq) + NH_3\ (aq) \rightarrow NH_4NO_3\ (aq)$$

(a)__________

(b)__________

3. A 10.0 mL sample of calcium hydroxide solution required 26.85 mL of 0.225 *M* hydrochloric acid for neutralization. Calculate (a) the molar concentration of the base.

$$2\ HCl(aq) + Ca(OH)_2\ (aq) \rightarrow CaCl_2(aq) + 2\ H_2O(l)$$

____

4. A Rolaids tablet contains calcium carbonate for neutralizing stomach acid. If a Rolaids tablet neutralizes 24.65 mL of 0.547 *M* hydrochloric acid, how many milligrams of calcium carbonate are in a Rolaids tablet?

$$CaCO_3(s) + 2\,HCl(aq) \rightarrow CaCl_2(aq) + H_2O(l) + CO_2(g)$$

____

5. (optional) A student diluted 15.0 mL of 6 *M* NaOH solution into 485.0 mL of distilled water. Calculate the molarity of the diluted base solution.

____

Explain why this diluted NaOH solution *cannot* be used as a standard solution.

EXPERIMENT

21

Electrical Conductivity of Aqueous Solutions

OBJECTIVES

- To observe the electrical conductivity of substances in aqueous solution.
- To determine whether an aqueous solution is a strong or weak electrolyte.
- To interpret a chemical reaction by observing aqueous solution conductivity.
- To become proficient in writing net ionic equations.

DISCUSSION

Electrical conductivity is based on the flow of electrons. Metals are good conductors of electricity because they allow electrons to flow through the metal. Distilled water is a very weak conductor because very little electricity passes through pure water. However, when a substance dissolves in water and forms ions, the ions are capable of conducting an electric current. If the substance is highly ionized, the solution is a strong conductor of electricity. If the substance is only slightly ionized, the solution is a weak conductor.

Soluble salts dissolve in water to form positive and negative ions. For example, sodium chloride dissolves in water to form Na^+ and Cl^- ions. The separation of ions in an ionic compound is termed **dissociation**.

A strong acid also dissolves in water to form positive and negative ions. For example, hydrogen chloride dissolves in water to form H^+ and Cl^- ions. The formation of positive and negative ions from a molecular compound, such as HCl, is termed **ionization**.

In this experiment, you will be testing conductivity using an apparatus that has two wires serving as electrodes (Figure 21.1). If the electrodes are immersed in a strong electrolyte solution, the circuit is completed and the light bulb in the apparatus glows brightly. If the electrodes are immersed in a weak electrolyte solution, the light bulb glows dimly.

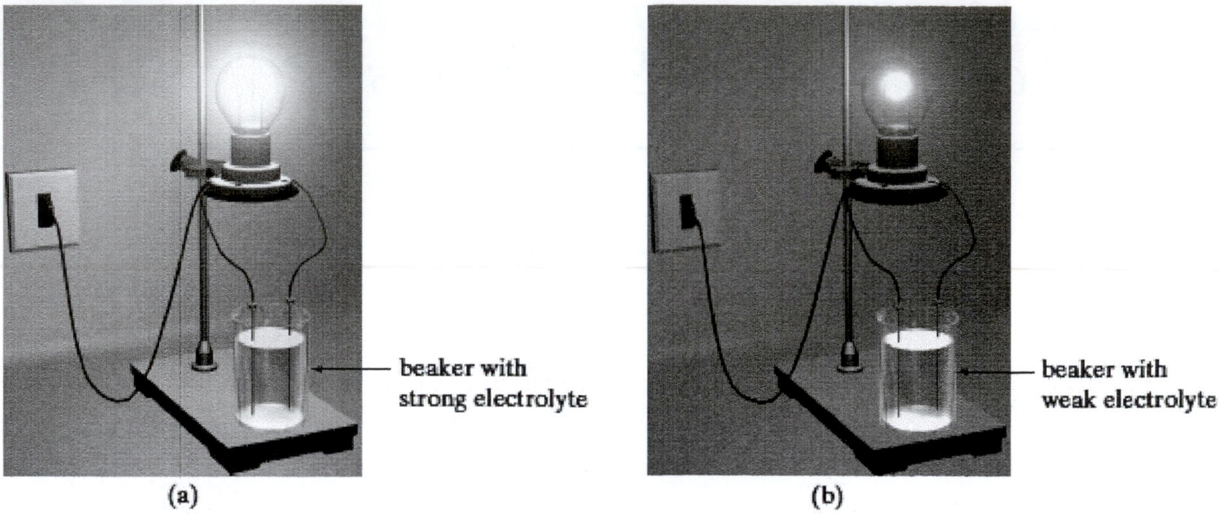

Figure 21.1 Apparatus for Conductivity Testing (a) A strong electrolyte is a good conductor of electricity and the light bulb glows brightly. (b) A weak electrolyte is a poor conductor of electricity and the light bulb glows dimly.

A solution that is a good conductor of an electric current is called a **strong electrolyte**. Examples of strong electrolytes include strong acids, strong bases, and salts that are highly soluble in aqueous solution. A solution that is a poor conductor of electricity is called a **weak electrolyte**. Examples of weak electrolytes include weak acids, weak bases, and salts that are only slightly soluble in aqueous solution. Table 21.1 lists several common examples of strong and weak electrolytes.

Table 21.1 Strong and Weak Electrolytes

Strong Electrolytes	Weak Electrolytes
Strong Acids	*Weak Acids*
hydrochloric acid, $HCl(aq)$	hydrofluoric acid, $HF(aq)$
nitric acid, $HNO_3(aq)$	acetic acid, $HC_2H_3O_2(aq)$
sulfuric acid, $H_2SO_4(aq)$	most other acids
Strong Bases	*Weak Bases*
sodium hydroxide, $NaOH(aq)$	ammonium hydroxide, $NH_4OH(aq)$
potassium hydroxide, $KOH(aq)$	most other bases
calcium hydroxide, $Ca(OH)_2(aq)$	
barium hydroxide, $Ba(OH)_2(aq)$	
Soluble Salts	*Very Slightly Soluble Salts*
sodium chloride, $NaCl(aq)$	silver chloride, $AgCl(s)$
sodium carbonate, $Na_2CO_3(aq)$	calcium carbonate, $CaCO_3(s)$
sodium sulfate, $Na_2SO_4(aq)$	barium sulfate, $BaSO_4(s)$

Since strong electrolytes are highly ionized, we will indicate these substances as ***ionized*** in aqueous solution. Conversely, since weak electrolytes are only slightly ionized, we will indicate these substances as ***nonionized*** in aqueous solution. The following examples illustrate writing strong and weak electrolytes in aqueous solution.

Example 21.1 • Ionization of a Strong Acid

Sulfuric acid is a strong acid and the light bulb gives a bright glow when tested by the conductivity apparatus. Write H_2SO_4 as it exists in aqueous solution.

Solution: Sulfuric acid is a strong electrolyte that is highly ionized. Thus, we will write aqueous H_2SO_4 as ***ionized:*** $2\ H^+(aq) + SO_4^{2-}(aq)$.

Example 21.2 • Ionization of a Weak Acid

Carbonic acid is a weak acid and the light bulb gives a dim glow when tested by the conductivity apparatus. Write H_2CO_3 as it exists in aqueous solution.

Solution: Carbonic acid is a weak electrolyte that is only slightly ionized. Thus, we will write aqueous H_2CO_3 as ***nonionized:*** $H_2CO_3(aq)$.

Example 21.3 • Ionization of a Strong Base

Potassium hydroxide is a strong base and the light bulb gives a bright glow when tested by the conductivity apparatus. Write KOH as it exists in aqueous solution.

Solution: Potassium hydroxide is a strong electrolyte that is highly ionized. Thus, we will write aqueous KOH as ***ionized:*** $K^+(aq) + OH^-(aq)$.

Example 21.4 • Ionization of a Weak Base

Ammonium hydroxide is a weak base and the light bulb gives a dim glow when tested by the conductivity apparatus. Write NH_4OH as it exists in aqueous solution.

Solution: Ammonium hydroxide is a weak electrolyte that is only slightly ionized. Thus, we will write aqueous NH_4OH as ***nonionized:*** $NH_4OH(aq)$.

Example 21.5 • Dissociation of a Soluble Salt

Aluminum chloride is a soluble salt and the light bulb gives a bright glow when tested by the conductivity apparatus. Write $AlCl_3$ as it exists in aqueous solution.

Solution: Aluminum chloride is a strong electrolyte that is highly ionized. Thus, we will write aqueous $AlCl_3$ as ***ionized:*** $Al^{3+}(aq) + 3\ Cl^-(aq)$.

Writing Net Ionic Equations

Given the chemical equation for a reaction, balance the equation by inspection. Next, convert the balanced chemical equation into a **net ionic equation**, using the following guidelines.

1. Write a substance in the chemical equation in the ionized form if it is a strong electrolyte. Examples of strong electrolytes include: strong acids, strong bases, and soluble salts. Refer to Table 21.1 for the most common examples.

2. Write a substance in the chemical equation in the nonionized form if it is a weak electrolyte. Examples of weak electrolytes include weak acids, weak bases, insoluble salts, and water.

3. Write the **total ionic equation** that shows highly ionized substances in the ionic form and weakly ionized substances in the nonionized form.

4. Convert the total ionic equation to a net ionic equation by canceling **spectator ions**. Spectator ions must be identical on both sides of the total ionic equation.

5. Check the net ionic equation for (a) mass balance, and (b) ionic charge balance.

Example 21.6 • Double Replacement Net Ionic Equation

$$CaCl_2(aq) + K_2CO_3(aq) \rightarrow CaCO_3(s) + 2\ KCl(aq)$$

$$Ca^{2+}(aq) + 2\ \cancel{Cl^-}(aq) + 2\ \cancel{K^+}(aq) + CO_3^{2-}(aq) \rightarrow CaCO_3(s) + 2\ \cancel{K^+}(aq) + 2\ \cancel{Cl^-}(aq)$$

$$Ca^{2+}(aq) + CO_3^{2-}(aq) \rightarrow CaCO_3(s)$$

Example 21.7 • Double Replacement Net Ionic Equation

$$2\ AlBr_3(aq) + 3\ MgCl_2(aq) \rightarrow 2\ AlCl_3(aq) + 3\ MgBr_2(aq)$$

$$2\ \cancel{Al^{3+}}(aq) + 6\ \cancel{Br^-}(aq) + 3\ \cancel{Mg^{2+}}(aq) + 6\ \cancel{Cl^-}(aq) \rightarrow$$

$$2\ \cancel{Al^{3+}}(aq) + 6\ \cancel{Cl^-}(aq) + 3\ \cancel{Mg^{2+}}(aq) + 6\ \cancel{Br^-}(aq)$$

All spectator ions; thus, all the ions cancel and there is *No Reaction (NR)*.

Example 21.8 • Neutralization Net Ionic Equation

$$H_2SO_4(aq) + 2\ NaOH(aq) \rightarrow Na_2SO_4(aq) + 2\ H_2O(l)$$

$$2\ H^+(aq) + \cancel{SO_4^{2-}}(aq) + 2\ \cancel{Na^+}(aq) + 2\ OH^-(aq) \rightarrow 2\ \cancel{Na^+}(aq) + \cancel{SO_4^{2-}}(aq) + 2\ H_2O(l)$$

$$H^+(aq) + OH^-(aq) \rightarrow H_2O(l)$$

EQUIPMENT and CHEMICALS

- conductivity apparatus
- small, dry beakers (6)
- glass stirring rod
- wash bottle with distilled water
- sodium chloride, solid NaCl
- calcium carbonate, solid $CaCO_3$
- calcium chloride, solid $CaCl_2$
- hydrochloric acid, 0.1 *M* HCl
- acetic acid, 0.1 *M* $HC_2H_3O_2$
- nitric acid, 0.1 *M* HNO_3
- sodium hydroxide, 0.1 *M* NaOH
- ammonium hydroxide, 0.1 *M* NH_4OH
- potassium iodide, 0.1 *M* KI
- aluminum nitrate, 0.1 *M* $Al(NO_3)_3$
- magnesium hydroxide, sat'd. $Mg(OH)_2$
- copper(II) sulfate, 0.1 *M* $CuSO_4$
- calcium nitrate, 0.1 *M* $Ca(NO_3)_2$
- sulfuric acid, 0.1 *M* H_2SO_4
- barium hydroxide, 0.1 *M* $Ba(OH)_2$
- straw

PROCEDURE

A. Conductivity Testing—Evidence for Ions in Aqueous Solution

Note: Rinse the electrodes with distilled water after each conductivity test. Record your observations in the Data Table, and state whether the conductivity test indicates a strong electrolyte or weak electrolyte. *(Write strong electrolytes as ions and weak electrolytes as molecules; refer to the LiOH and* HNO_2 *examples in the Data Table.)*

1. Pour about 25 mL of *distilled* water in a small dry beaker and test the conductivity. Pour about 25 mL of *tap* water in a small dry beaker and test the conductivity.

2. Place about 0.5 g of solid NaCl in a small dry beaker and test the conductivity. Add distilled water, stir, and retest the conductivity.

3. Place about 0.5 g of solid $CaCO_3$ in a small dry beaker and test the conductivity. Add distilled water, stir, and retest the conductivity.

4. Place about 0.5 g of solid $CaCl_2$ in a small dry beaker and test the conductivity. Add distilled water, stir, and retest the conductivity.

5. Test the conductivity of each of the following in a small beaker:
   (a) ~10 mL hydrochloric acid, 0.1 *M* HCl
   (b) ~10 mL acetic acid, 0.1 *M* $HC_2H_3O_2$
   (c) ~10 mL nitric acid, 0.1 *M* HNO_3
   (d) ~10 mL sodium hydroxide, 0.1 *M* NaOH
   (e) ~10 mL ammonium hydroxide, 0.1 *M* NH_4OH
   (f) ~10 mL potassium iodide, 0.1 *M* KI
   (g) ~10 mL aluminum nitrate, 0.1 *M* $Al(NO_3)_3$
   (h) ~10 mL magnesium hydroxide, saturated $Mg(OH)_2$
   (i) ~10 mL copper(II) sulfate, 0.1 *M* $CuSO_4$
   (j) ~10 mL calcium nitrate, 0.1 *M* $Ca(NO_3)_2$

RECYCLE
Chemical
Waste

B. Conductivity Testing—Evidence for a Chemical Reaction

1. Test the conductivity of 0.1 *M* $HC_2H_3O_2$ and 0.1 *M* NH_4OH in separate beakers. Pour the solutions together and retest the conductivity. Record your observations and conclusions in the Data Table. Balance the equation for the reaction, and write the total ionic and net ionic equations.

2. Test the conductivity of 0.1 *M* H_2SO_4 and 0.1 *M* $Ba(OH)_2$ in separate beakers. Add 10 drops of 0.1 *M* H_2SO_4 into a beaker containing ~25 mL of distilled water. Continuously test the conductivity while adding 0.1 *M* $Ba(OH)_2$ dropwise until the conductivity is minimal. Record your observations and conclusions. Balance the equation for the reaction, and write the total ionic and net ionic equations.

RECYCLE
Chemical
Waste

3. Test the conductivity of distilled water while blowing CO_2 through a straw in the water. Add 10 drops of 0.1 *M* $Ba(OH)_2$ into a beaker containing ~25 mL of distilled water. Continuously test the conductivity while blowing through a straw into the solution until the conductivity is minimal. Record your observations and conclusions. Balance the equation for the reaction, and write the total ionic and net ionic equations.

C. Net Ionic Equations—A Study Assignment

Balance the following neutralization reactions; write the total ionic and net ionic equations. Refer to *Writing Net Ionic Equations* for directions and examples.

1. Strong Acid and Strong Base:

$$HCl(aq) \quad + \quad NaOH(aq) \quad \rightarrow \quad NaCl(aq) \quad + \quad H_2O(l)$$

2. Strong Acid and Weak Base:

$$HCl(aq) \quad + \quad NH_4OH\,(aq) \quad \rightarrow \quad NH_4Cl(aq) \quad + \quad H_2O(l)$$

3. Weak Acid and Strong Base:

$$HF(aq) \quad + \quad NaOH(aq) \quad \rightarrow \quad NaF(aq) \quad + \quad H_2O(l)$$

4. Weak Acid and Weak Base:

$$HF(aq) \quad + \quad NH_4OH\,(aq) \quad \rightarrow \quad NH_4F(aq) \quad + \quad H_2O(l)$$

EXPERIMENT 21 NAME ______________________

DATE ______________ SECTION ______________________

PRELABORATORY ASSIGNMENT*

1. Provide the key term that corresponds to each of the following definitions.

__________ (a) an aqueous solution that is a good conductor of electricity and produces a bright glow from a light bulb in a conductivity apparatus

__________ (b) an aqueous solution that is a poor conductor of electricity and produces a dim glow from a light bulb in a conductivity apparatus

__________ (c) the process of an ionic compound dissolving in water and separating into positive and negative ions

__________ (d) the process of a polar molecular compound dissolving in water and forming positive and negative ions

__________ (e) a chemical equation that portrays highly ionized substances in the ionic form and slightly ionized substances in the nonionized form

__________ (f) ions in aqueous solution that do not participate in a reaction, and do not appear in the net ionic equation

__________ (g) a chemical equation that portrays an ionic reaction after spectator ions have been canceled from the total ionic equation

Key Terms: dissociation, ionization, net ionic equation, spectator ions, strong electrolyte, total ionic equation, weak electrolyte

2. What is observed when conductivity testing the following:

(a) strong electrolyte

(b) weak electrolyte

3. Refer to Table 21.1 and indicate whether the following are strong or weak electrolytes.

(a) $HCl(aq)$ (b) $HC_2H_3O_2(aq)$

(c) $NaOH(aq)$ (d) $NH_4OH(aq)$

(e) $NaCl(aq)$ (f) $AgCl(s)$

---

* *Answers in Appendix J*

4. Distinguish between $NaCl(s)$ and $NaCl(aq)$.

5. Given the following observations, write each of the following as it exists in aqueous solution.

   (a) $HI(aq)$ — strong electrolyte

   (b) $HF(aq)$ — weak electrolyte

   (c) $Sr(OH)_2(aq)$ — strong electrolyte

   (d) $AgNO_3(aq)$ — strong electrolyte

   (e) $Ag_2SO_4(s)$ — weak electrolyte

6. Why must the electrodes on the conductivity apparatus, as well as all beakers, be rinsed with *distilled* water before each conductivity test?

7. What safety precautions should be observed in this experiment?

EXPERIMENT 21 NAME ______________________________

DATE ____________________ SECTION ______________________________

DATA TABLE

A. Conductivity Testing—Evidence for Ions in Aqueous Solution

Solution	*Observation*	*Conclusion*	*Ionized/Nonionized*
LiOH (aq)	bulb glows brightly	strong electrolyte	$Li^{+}(aq)$ + $OH^{-}(aq)$
$HNO_2(aq)$	bulb glows dimly	weak electrolyte	$HNO_2(aq)$
1. H_2O – distilled			
H_2O – tap			— *omit* —
2. NaCl (s)			
NaCl (aq)			
3. $CaCO_3(s)$			
$CaCO_3(aq)$			
4. $CaCl_2(s)$			
$CaCl_2(aq)$			

RECYCLE
Chemical
Waste

Solution	*Observation*	*Conclusion*	*Ionized/Nonionized*
5. (a) $HCl(aq)$			
(b) $HC_2H_3O_2(aq)$			
(c) $HNO_3(aq)$			
(d) $NaOH\ (aq)$			
(e) $NH_4OH(aq)$			
(f) $KI\ (aq)$			
(g) $Al(NO_3)_3(aq)$			
(h) $Mg(OH)_2(aq)$			
(i) $CuSO_4(aq)$			
(j) $Ca(NO_3)_2(aq)$			

RECYCLE
Chemical
Waste

B. Conductivity Testing—Evidence for a Chemical Reaction

Solution	*Observation*	*Conclusion*
1. (a) $HC_2H_3O_2(aq)$	______________	______________
(b) $NH_4OH(aq)$	______________	______________
(c) $HC_2H_3O_2(aq)$ + $NH_4OH(aq)$	______________	______________

equation: $HC_2H_3O_2(aq) + NH_4OH(aq) \rightarrow NH_4C_2H_3O_2(aq) + H_2O(l)$

total ionic:

net ionic:

Solution	*Observation*	*Conclusion*
2. (a) $H_2SO_4(aq)$	______________	______________
(b) $Ba(OH)_2(aq)$	______________	______________
(c) $H_2SO_4(aq)$ + $Ba(OH)_2(aq)$	______________	______________

RECYCLE
Chemical
Waste

equation: $H_2SO_4(aq) + Ba(OH)_2(aq) \rightarrow BaSO_4(s) + H_2O(l)$

total ionic:

net ionic:

Solution	*Observation*	*Conclusion*
3. (a) $Ba(OH)_2(aq)$ — see 2. (b) above	______________	______________
(b) $CO_2(g)$	______________	______________
(c) $Ba(OH)_2(aq)$ + $CO_2(g)$	______________	______________

equation: $Ba(OH)_2(aq) + CO_2(g) \rightarrow BaCO_3(s) + H_2O(l)$

total ionic:

net ionic:

C. Net Ionic Equations—A Study Assignment

1. Strong Acid and Strong Base:

equation: $HCl(aq)$ + $NaOH(aq)$ → $NaCl(aq)$ + $H_2O(l)$

total ionic:

net ionic:

---

2. Strong Acid and Weak Base:

equation: $HCl(aq)$ + $NH_4OH(aq$ → $NH_4Cl(aq)$ + $H_2O(l)$

total ionic:

net ionic:

---

3. Weak Acid and Strong Base:

equation: $HF(aq)$ + $NaOH(aq)$ → $NaF(aq)$ + $H_2O(l)$

total ionic:

net ionic:

---

4. Weak Acid and Weak Base:

equation: $HF(aq)$ + $NH_4OH(aq$ → $NH_4F(aq)$ + $H_2O(l)$

total ionic:

net ionic:

EXPERIMENT 21 NAME ______________________

DATE ______________ SECTION ______________________

POSTLABORATORY ASSIGNMENT

1. Why is distilled water a weaker conductor than tap water? (Refer to **A.1** in the Data Table.)

____

2. Why is solid sodium chloride a weak electrolyte, while aqueous NaCl is a strong electrolyte? (Refer to **A.2** in the Data Table.)

____

3. Why is calcium carbonate a weak electrolyte, while calcium chloride is a strong electrolyte? (Refer to **A.3** and **A.4** in the Data Table.)

____

4. Why are aqueous solutions of $HC_2H_3O_2$ and NH_4OH weak electrolytes individually, but a strong electrolyte after they are added together? (Refer to **B.1** in the Data Table.)

____

5. Why are aqueous solutions of H_2SO_4 and $Ba(OH)_2$ strong electrolytes individually, but a weak electrolyte after they are added together? (Refer to **B.2** in the Data Table.)

____

6. Why does blowing carbon dioxide gas into aqueous barium hydroxide, $Ba(OH)_2$, reduce the conductivity from a strong electrolyte to a weak electrolyte? (Refer to **B.3** in the Data Table.)

____

7. Balance the following precipitation reactions; write the total ionic and net ionic equations.

(a) $CuCl_2(aq) + K_2CO_3(aq) \rightarrow CuCO_3(s) + KCl(aq)$

total ionic:

net ionic:

(b) $CuCl_2(aq) + Na_2S(aq) \rightarrow CuS(s) + NaCl(aq)$

total ionic:

net ionic:

(c) $CuCl_2(aq) + NH_4OH(aq) \rightarrow Cu(OH)_2(s) + NH_4Cl(aq)$

total ionic:

net ionic:

8. (optional) Write a balanced chemical equation for the addition of aqueous calcium chloride and lithium nitrate solutions. Show the total ionic and net ionic equations.

EXPERIMENT

22

Activity Series for Metals

OBJECTIVES

- To observe the oxidation numbers for an element in a compound or ion.
- To write balanced chemical equations for redox reactions.
- To place an unknown metal (**X**) in an activity series.

DISCUSSION

An **oxidation number** is the positive or negative number that describes the combining capacity of an element in a compound or polyatomic ion. Elements that are not combined with another element are assigned a value of zero. For example, iron metal and oxygen gas each have a value of zero. However, in ferrous oxide (FeO), the ferrous ion (Fe^{2+}) has an oxidation number of +2 and the oxide ion (O^{2-}) has an oxidation number of –2. The following examples illustrate the calculation of oxidation numbers.

Example Exercise 22.1 • Oxidation Number of C in a Compound

Calculate the oxidation number of carbon in sodium hydrogen carbonate, $NaHCO_3$.

Solution: The oxidation number (ox no) of sodium ion is +1, of a hydrogen ion, +1, and that of oxygen, –2. Since the sum of all oxidation numbers for the elements in a compound must equal zero, we can write the following equation.

$$+1 +1 + \text{ox no C} + 3(-2) = 0$$

$$+2 + \text{ox no C} - 6 = 0$$

$$\text{ox no C} = +4$$

Balancing Redox Equations

A process in which a substance undergoes an increase in oxidation number by losing electrons is called **oxidation**; a decrease in oxidation number by gaining electrons is called **reduction**. To balance an oxidation-reduction reaction, that is, a **redox reaction**, we can either use the *oxidation number method,* or the *half-reaction method.* The following exercises illustrate.

Example Exercise 22.2 • Balancing Equations by Oxidation Numbers

Balance the following redox reaction in nitric acid by the *oxidation number* method.

$$Ag + NO_3^- \rightarrow Ag^+ + NO \quad (\textit{in acid})$$

Solution: First, note the change in oxidation number for silver: 0 to +1. Silver loses 1e–. Second, note the change in oxidation number for nitrogen: +5 to +2. Nitrogen gains 3e–. To balance electrons, the coefficient of silver is three; thus, 3 Ag and 3 Ag^+. The coefficient of each nitrogen species is one; thus,

$$3\,Ag + NO_3^- \rightarrow 3\,Ag^+ + NO$$

The next step is to balance the oxygen atoms, using water molecules.

$$3\,Ag + NO_3^- \rightarrow Ag^+ + NO + 2\,H_2O$$

Since the reaction takes place in acidic solution, we will balance the hydrogen atoms using hydrogen ions.

$$4\,H^+ + 3\,Ag + NO_3^- \rightarrow 3\,Ag^+ + NO + 2\,H_2O$$

The last step is to check the equation for balance. A redox equation must be balanced in terms of the number of atoms. Also, you should check that the total charge of the reactants equals that of the products.

$$4\,H^+ + 3\,Ag + NO_3^- \rightarrow 3\,Ag^+ + NO + 2\,H_2O$$

The check reveals that the atoms of reactants and products are balanced. Furthermore, the total ionic charge of the reactants is 3+; the total product charge is also 3+.

Example Exercise 22.3 • Balancing Redox Equations by Half-Reactions

Balance the following redox reaction in nitric acid by the *half-reaction* method.

$$Ag + NO_3^- \rightarrow Ag^+ + NO \quad \textit{(in acid)}$$

Solution: Using this method, the oxidation half-reaction is treated separately from the reduction half-reaction. The two half-reactions are then added together to give a balanced redox equation. First, write the equation for each half-reaction.

$$Ag \rightarrow Ag^+$$
$$NO_3^- \rightarrow NO$$

Balance the partial equation, using water and hydrogen ions.

$$Ag \rightarrow Ag^+$$
$$4\ H^+ + NO_3^- \rightarrow NO + 2\ H_2O$$

Next, write the two half-reactions, and balance the charge using electrons.

oxidation: $Ag \rightarrow Ag^+ + e^-$

reduction: $4\ H^+ + NO_3^- + 3\ e^- \rightarrow NO + 2\ H_2O$

The reduction half-reaction gains 3e– for every 1e– lost in the oxidation half-reaction. Therefore, we will multiply the oxidation half-reaction by three.

$$3\ Ag \rightarrow 3\ Ag^+ + 3\ e^-$$

Let's add the two half-reactions together to obtain the overall redox equation. After canceling 3 e– from each side of the equation, we have

$$4\ H^+ + 3\ Ag + NO_3^- \rightarrow 3\ Ag^+ + NO + 2\ H_2O$$

The number of atoms of reactants and products are equal. Also notice that the total ionic charge is 3+ for both reactants and products.

Determining an Activity Series

An **activity series** is a list of metals arranged in order of their ability to displace another metal from aqueous solution. A metal that is more active will reduce another metal ion to a metal in the free state. For example, zinc metal is more active than copper metal and can reduce Cu^{2+} to Cu.

$$Zn(s) + Cu^{2+}(aq) \rightarrow Zn^{2+}(aq) + Cu(s)$$

The reverse process gives no reaction. That is, copper metal cannot reduce Zn^{2+} to Zn, because zinc is less active than copper, according to the activity series.

$$Cu(s) + Zn^{2+}(aq) \rightarrow NR$$

For reference purposes, hydrogen (H) is included in the series. Metals above hydrogen in the activity series displace hydrogen gas from acid solutions. Metals below hydrogen do not react with dilute acids. Example Exercise 22.5 illustrates the determination of an activity series.

Example Exercise 22.4 • Determining an Activity Series

Determine the activity series for zinc, iron, copper, and hydrogen, based on the following observations.

1. Zinc metal obtains a dull gray appearance when placed in an aqueous iron solution. The equation for the reactions is:

$$Zn(s) \quad + \quad Fe^{2+}(aq) \quad \rightarrow \quad Zn^{2+}(aq) \quad + \quad Fe(s)$$

2. An iron nail releases gas bubbles in dilute acid according to the equation:

$$Fe(s) \quad + \quad 2\ H^{+}(aq) \quad \rightarrow \quad Fe^{2+}(aq) \quad + \quad H_2(g)$$

3. Copper metal does not react with dilute acid. Thus,

$$Cu(s) \quad + \quad H^{+}(aq) \quad \rightarrow \quad NR$$

Solution: Since zinc metal displaces iron from aqueous solution, we conclude that Zn is more active than Fe (Zn > Fe).

Since the iron nail releases hydrogen gas from an aqueous acid, we conclude that Fe is more active than H (Fe > H).

Since copper metal does not react with an aqueous acid, we conclude that Cu is less active than hydrogen (H > Cu).

In summary, we conclude that the activity series is

$$Zn \quad > \quad Fe \quad > \quad (H) \quad > \quad Cu$$

most active *least active*

EQUIPMENT and CHEMICALS

- 13 x 100 mm test tubes (6) & test tube rack
- test tube brush
- wash bottle with distilled water
- evaporating dish
- iron wire, Fe
- ferric oxide, powder, Fe_2O_3
- ferric nitrate, 0.5 *M* $Fe(NO_3)_3$
- stannous chloride, 0.5 *M* $SnCl_2$
- manganese chips, Mn
- potassium permanganate, 0.010 *M* $KMnO_4$
- dilute sodium hydroxide, 6 *M* NaOH
- sodium sulfite, 0.5 *M* Na_2SO_3
- dilute hydrochloric acid, 6 *M* HCl
- sulfur, S powder
- sodium sulfide, solid Na_2S
- dilute sulfuric acid, 3 *M* H_2SO_4
- iodine solution, 0.5 *M* I_2/KI
- sodium thiosulfate, 0.1 *M* $Na_2S_2O_3$
- dilute nitric acid, 6 *M* HNO_3
- copper, Cu metal
- concentrated nitric acid, 16 *M* HNO_3
- ammonium chloride, solid NH_4Cl
- sodium sulfite, 0.5 *M* Na_2SO_3
- magnesium, Mg metal
- zinc, Zn metal
- zinc sulfate, 0.1 *M* $ZnSO_4$
- silver nitrate, 0.1 *M* $AgNO_3$
- unknown metal samples (**X**)

PROCEDURE

A. Oxidation Numbers of Iron

1. Inspect a piece of iron wire. Record your observations in the Data Table, and state the oxidation number for iron wire, Fe.

2. Inspect a sample of ferric oxide powder. Record your observations in the Data Table, and calculate the oxidation number for iron in Fe_2O_3.

3. Put 1 mL of aqueous ferric nitrate into a test tube. Note the color, and calculate the oxidation number for iron in $Fe(NO_3)_3$. Add 1 mL of aqueous stannous chloride to the test tube, and observe the color change as the ferrous ion is formed in solution. Record your observations, and state the oxidation number for iron in Fe^{2+}.

B. Oxidation Numbers of Manganese

1. Examine a piece of manganese metal. Record your observations in the Data Table, and state the oxidation number for manganese metal, Mn.

2. Place 2 mL of potassium permanganate in a test tube. Add a drop of dilute hydrochloric acid and several drops of sodium sulfite to produce the manganese(II) ion. Record your observations, and state the oxidation number for manganese in Mn^{2+}.

RECYCLE
Chemical Waste

3. Introduce 2 mL of potassium permanganate solution into a test tube. Add sodium sulfite drop by drop until the purple color fades. After several minutes, observe the solid particles of manganese dioxide forming in the solution. Record your observations, and calculate the oxidation number for manganese in MnO_2.

4. Deliver 2 mL of potassium permanganate into a test tube. Note the color, and calculate the oxidation number for manganese in the permanganate ion, MnO_4^-. Add a few drops of dilute sodium hydroxide and one drop of sodium sulfite into the test tube. Observe the color change and the formation of the manganate ion. Calculate the oxidation number for manganese in the manganate ion, MnO_4^{2-}.

C. Oxidation Numbers of Sulfur

1. Examine a small portion of powdered sulfur. Record your observations in the Data Table, and state the oxidation number for sulfur, S.

2. Place a pea-sized portion of sulfur in an evaporating dish, and ignite it with a burner. Notice the color of the flame and the white sulfur dioxide gas. Calculate the oxidation number for sulfur in gaseous SO_2.

   Caution: Perform this operation under a fume hood. Avoid breathing the sulfur dioxide gas, and dispose of the contents in the waste container.

3. Put a very small crystal of sodium sulfide, Na_2S, in a dry test tube. Add several drops of dilute sulfuric acid, H_2SO_4. Describe the odor of the hydrogen sulfide, H_2S, gas given off. Calculate the oxidation number for sulfur in Na_2S, H_2SO_4, and H_2S.

   Caution: Perform this operation under a fume hood. Avoid breathing the hydrogen sulfide gas, and dispose of the contents in the waste container.

4. Pour about 2 mL of iodine solution into a test tube. Add sodium thiosulfate, $Na_2S_2O_3$, until the iodine is discolored. Record the change, and calculate the oxidation number for sulfur in $Na_2S_2O_3$.

D. Oxidation Numbers of Nitrogen

1. Air contains over 78% nitrogen. By simply breathing air, note the odor and observe the color. State the oxidation number for nitrogen, N_2, in air.

2. Put 2 mL of *dilute* nitric acid, HNO_3, in a test tube. Add a piece of copper metal, and observe the reaction that releases nitrogen monoxide, NO. Record your observations, and calculate the oxidation number for nitrogen in HNO_3 and NO.

3. Deliver 2 mL of *concentrated* nitric acid, HNO_3, into a test tube. Add a piece of copper metal, and observe the reaction that evolves nitrogen dioxide gas, NO_2. Record your observations, and calculate the oxidation number for nitrogen in NO_2.

RECYCLE
Chemical
Waste

Caution: Perform this operation under a fume hood. Concentrated HNO_3 should be handled with caution. Avoid breathing the fumes, and dispose of the test tube contents in the waste container.

4. Using a spatula, introduce a pea-sized portion of solid ammonium chloride, NH_4Cl, into a test tube. Add a dropper of dilute sodium hydroxide, and cover the end of the test tube with your thumb for 30 seconds. Shake the test tube, release your thumb, and waft the gas to detect the odor. Calculate the oxidation number for nitrogen in ammonium chloride, NH_4Cl, and ammonia gas, NH_3.

E. Oxidation-Reduction Equations – A Study Assignment

Many of the reactions in Procedures A–D illustrate oxidation-reduction and are listed in the Data Table. Balance each redox reaction using either the *oxidation number* method, or the *half-reaction* method.

F. Activity Series and an Unknown Metal

1. Obtain an unknown metal (**X**), and record the number in the Data Table.

2. Add 2 mL of dilute hydrochloric acid to each of four test tubes. Drop a small piece of Cu, Mg, Zn, or **X** into each test tube. Record your observations in the Data Table.

3. Clean the test tubes, and add 2 mL of zinc sulfate solution. Put a small piece of Cu, Mg, Zn, or **X** into each test tube. Record your observations in the Data Table.

4. Clean the four test tubes, and then add 2 mL of silver nitrate solution. Put a small piece of Cu, Mg, Zn, or **X** into each test tube. Record your observations.

5. Based on the foregoing observations, list the experimental activity series for the following: Cu, Mg, Zn, Ag, (H), and **X**.

EXPERIMENT 22 NAME ____________________

DATE ____________ SECTION ____________________

*PRELABORATORY ASSIGNMENT**

1. Provide the key term that corresponds to each of the following definitions.

____________ (a) a chemical reaction that involves electron transfer and causes reduction of one species and oxidation of another

____________ (b) a chemical process characterized by the loss of electrons

____________ (c) a chemical process characterized by the gain of electrons

____________ (d) a substance that causes the oxidation of another substance in a redox reaction

____________ (e) a substance that causes the reduction of another substance in a redox reaction

____________ (f) a positive or negative value assigned to an atom in a substance according to a set of rules

____________ (g) a relative order of metals arranged in a list according to their ability to undergo reaction

Key Terms: activity series, oxidation, oxidation number, oxidizing agent, redox reaction, reducing agent, reduction

2. Calculate the oxidation number of chlorine in each of the following.

(a) Cl_2 ____________

(b) HCl ____________

(c) Cl_2O_7 ____________

(d) $KClO$ ____________

(e) ClO_3^- ____________

* *Answers in Appendix J*

3. Balance the following redox reaction, which takes place in an acidic solution.

$$Cr_2O_7^{2-} + Fe^{2+} + H^+ \rightarrow Cr^{3+} + Fe^{3+} + H_2O$$

4. Balance the following redox reaction, which takes place in a basic solution.

$$Cr_2O_7^{2-} + Fe^{2+} + OH^- \rightarrow Cr^{3+} + Fe^{3+} + H_2O$$

5. The following reactions take place in aqueous solution:

$$Mg + Mn^{2+} \rightarrow Mg2^+ + Mn$$

$$Mn + Ni^{2+} \rightarrow Mn^{2+} + Ni$$

$$Ni + Ag^+ \rightarrow Ni^{2+} + Ag$$

Write the activity series for the metals Mg, Mn, Ni, and Ag. List the most active metal first.

6. What safety precautions must be observed in this experiment?

EXPERIMENT 22 NAME ______________________

DATE ______________ SECTION ______________________

DATA TABLE

A. Oxidation Numbers of Iron

Substance	*Observation*	*Oxidation Number*
1. Fe		________
2. Fe_2O_3		________
3. $Fe(NO_3)_3$		________
Fe^{2+}		________

RECYCLE Chemical Waste

B. Oxidation Numbers of Manganese

1. Mn		________
2. Mn^{2+}		________
3. MnO_2		________
4. MnO_4^-		________
MnO_4^{2-}		________

C. Oxidation Numbers of Sulfur

1. S ________

2. SO_2 ________

3. Na_2S ________

   H_2SO_4 ________

   H_2S ________

4. $Na_2S_2O_3$ ________

RECYCLE
Chemical
Waste

D. Oxidation Numbers of Nitrogen

1. N_2 ________

2. HNO_3 ________

   NO ________

3. NO_2 ________

4. NH_4Cl ________

   NH_3 ________

E. Oxidation–Reduction Equations – A Study Assignment

1. Reduction of Fe^{3+} in neutral solution

$$Fe^{3+} + Sn^{2+} \rightarrow Fe^{2+} + Sn^{4+}$$

2. Reduction of MnO_4^- in acidic solution

$$MnO_4^- + SO_3^{2-} \rightarrow Mn^{2+} + SO_4^{2-}$$

3. Reduction of MnO_4^- in neutral solution

$$MnO_4^- + SO_3^{2-} \rightarrow MnO_2 + SO_4^{2-}$$

4. Reduction of MnO_4^- in basic solution

$$MnO_4^- + SO_3^{2-} \rightarrow MnO_4^{2-} + SO_4^{2-}$$

5. Oxidation of $S_2O_3^{2-}$ in neutral solution

$$S_2O_3^{2-} + I_2 \rightarrow S_4O_6^{2-} + I^-$$

6. Oxidation of Cu in dilute acid

$$Cu + NO_3^- \rightarrow Cu^{2+} + NO$$

7. Oxidation of Cu in concentrated acid

$$Cu + NO_3^- \rightarrow Cu^{2+} + NO_2$$

F. Activity Series and an Unknown Metal **UNKNOWN #** ______

	Cu	Mg	Zn	**X**
HCl				
$ZnSO_4$				
$AgNO_3$				

RECYCLE
Chemical
Waste

1. Based on the reactions with hydrochloric acid, which metals are more active than hydrogen (H)?

2. Based on the reactions with zinc sulfate, which metals are more active than zinc?

3. Based on the reactions with silver nitrate, which metals are more active than silver?

Activity Series for Cu, Mg, Zn, Ag, (H), and ***X****:*

______ ______ ______ ______ ______ ______

most active *least active*

EXPERIMENT 22 NAME ____________________

DATE ____________ SECTION ____________________

POSTLABORATORY ASSIGNMENT

1. Calculate the oxidation number for iodine in each of the following.

   (a) I_2 ____________

   (b) IF_7 ____________

   (c) I_2O_5 ____________

   (d) KIO_2 ____________

   (e) IO^- ____________

2. Balance the following redox reactions, which take place in an acidic solution.

   (a) $MnO_4^- \quad + \quad Fe^{2+} \quad \rightarrow \quad Mn^{2+} \quad + \quad Fe^{3+}$

   (b) $CrO_4^{2-} \quad + \quad C_2O_4^{2-} \quad \rightarrow \quad Cr^{3+} \quad + \quad CO_2$

3. Balance the following redox reactions, which take place in a basic solution.

(a) SO_3^{2-} + Cl_2 → SO_4^{2-} + Cl^-

(b) MnO_4^- + ClO_2^- → MnO_2 + ClO_3^-

4. Bromine water was added dropwise to test tubes containing aqueous solutions of bromide, chloride, and iodide. Iodide was oxidized to iodine; bromide and chloride gave no reaction. Arrange Br_2, Cl_2, I_2 in order of their strength as an oxidizing agent.

Strongest oxidizing agent: __________ __________ __________

5. (optional) Consider the following redox equation:

$$2\,KI + Cl_2 \rightarrow I_2 + 2\,KCl$$

Identify the reactant undergoing:

(a) oxidation __________ (b) reduction __________

Identify the reactant that is the:

(a) oxidizing agent __________ (b) reducing agent __________

Organic Models and Classes of Compounds

EXPERIMENT

23

OBJECTIVES

- To build molecular models for the following hydrocarbons: alkanes, alkenes, alkynes, and arenes.
- To build molecular models for the following hydrocarbon derivatives: organic halides, alcohols, phenols, ethers, amines, aldehydes, ketones, carboxylic acids, esters, and amides.
- To identify the class of compound for unknown molecular models.

DISCUSSION

Organic chemistry is the study of compounds that contain carbon. Inorganic compounds do not contain carbon. Interestingly, the element carbon is found in over 10 million different compounds. There are two reasons why over 90% of all compounds are organic. First, carbon is unusual in that it has the ability to self-link, forming chains of carbon atoms. Second, organic compounds typically contain several carbon atoms that may form different structures by joining together in more than one arrangement.

Compounds having the same molecular formula but different structural formulas are called **isomers**. For instance, the molecular formula C_4H_{10} may be constructed in two ways and will satisfy the bond requirements for carbon (four bonds) and hydrogen (one bond). Figure 23.1 illustrates the isomers of butane, C_4H_{10}. Although the entire molecule, as well as the individual bonds, can be rotated in space to give what appears to be additional structures for C_4H_{10}, careful examination will reveal that there are only two possibilities.

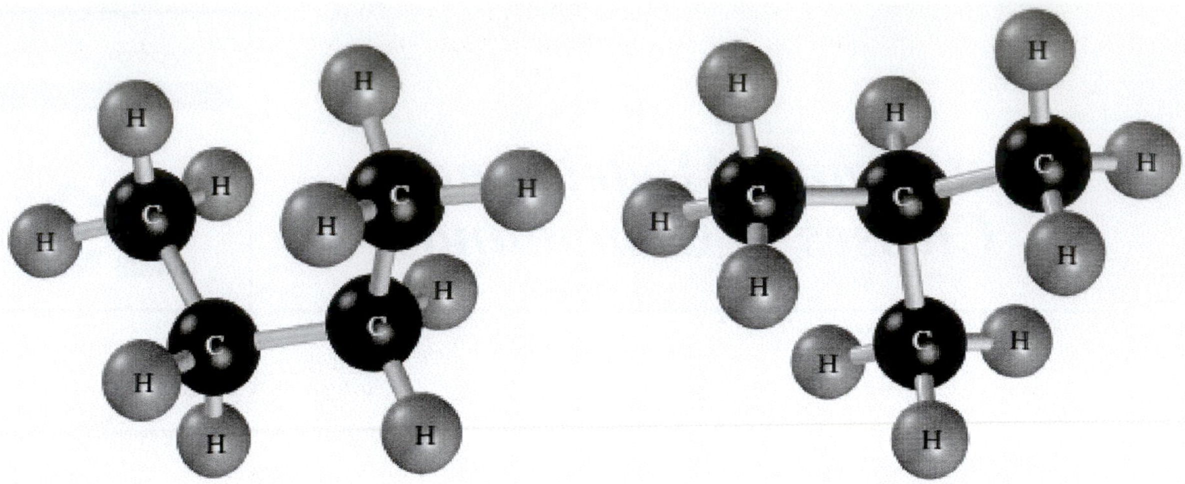

Figure 23.1 Structural Isomers The two isomers have the same formula, C_4H_{10}. Butane is shown on the left and "isobutane" is shown on the right.

There are millions of organic compounds that may be classified into only a few families. A **hydrocarbon** contains hydrogen and carbon, and may be classified as an alkane, alkene, alkyne, or arene. A **hydrocarbon derivative** is derived from a hydrocarbon, and often contains oxygen, nitrogen, or a halogen. A **functional group** characterizes each **class of compounds**, and imparts similar properties to each member of the family. Figure 23.2 illustrates the classification for organic compounds into four hydrocarbon classes and ten hydrocarbon derivatives.

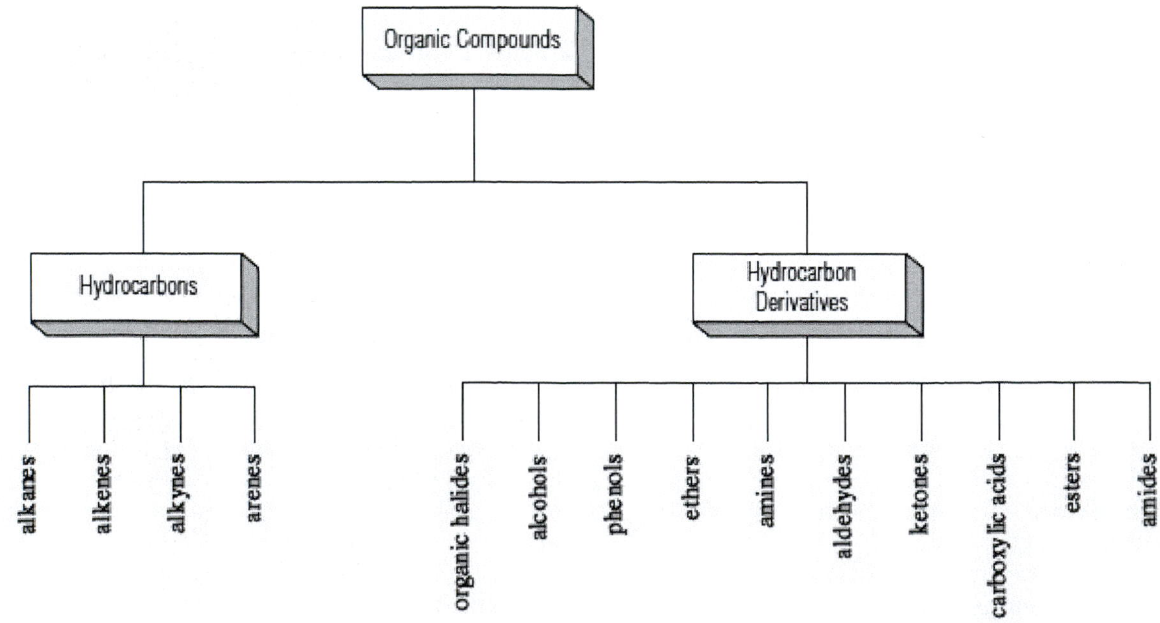

Figure 23.2 Classes of Organic Compounds Organic compounds are organized systematically into four classes of hydrocarbons, and ten classes of hydrocarbon derivatives as shown above.

In this experiment, we will build example molecular models for each of the hydrocarbons and derivatives shown in Figure 23.2. The following examples will serve to correlate the classes of compounds with the molecular models you will construct.

Class of Compound	*Example*	*Model Representation*
alkane	CH_3-CH_3	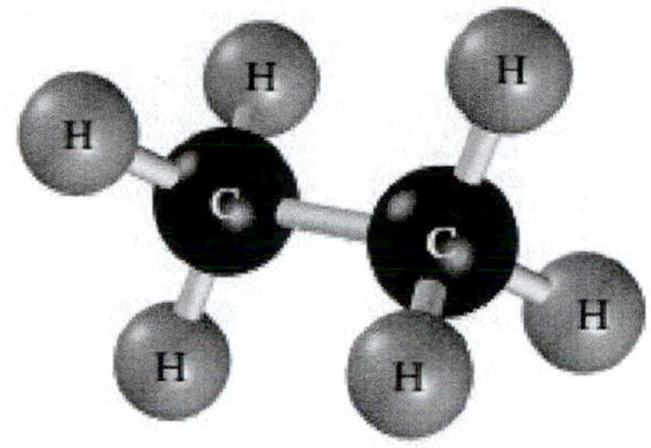
alkene	$CH_2=CH_2$	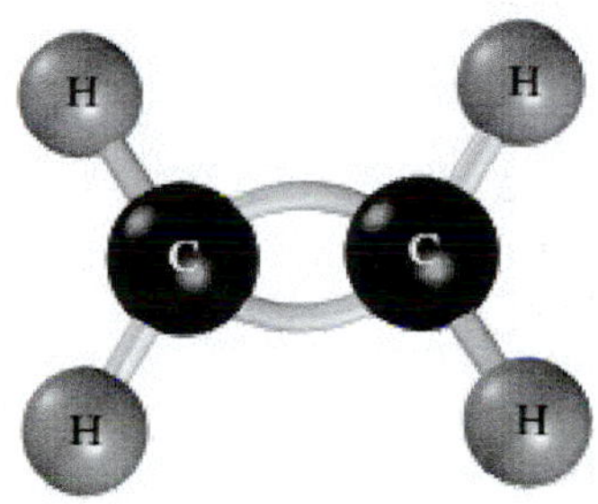
alkyne	$HC\equiv CH$	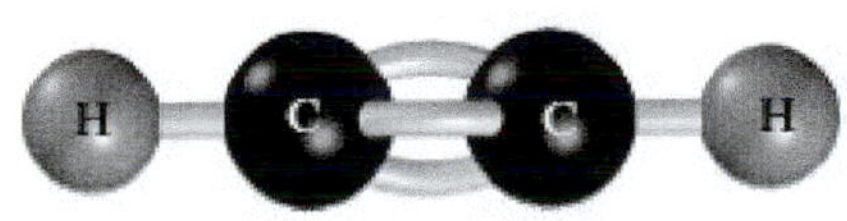
arene	C_6H_6	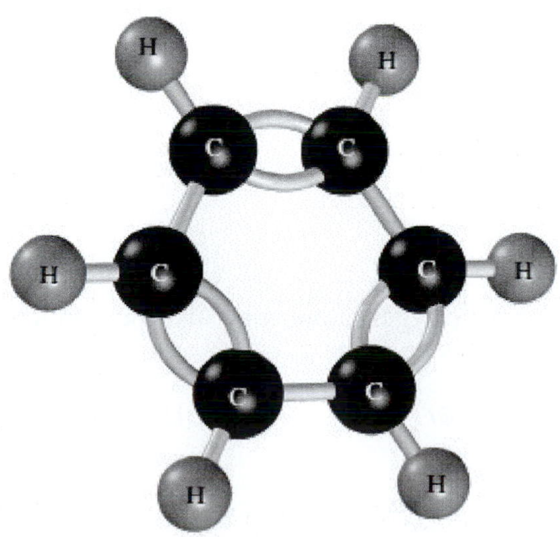

Substituting a chlorine, bromine, or iodine atom for a hydrogen atom onto the hydrocarbon chain produces a class of compounds called the *organic halides*.

organic halide CH_3-Cl

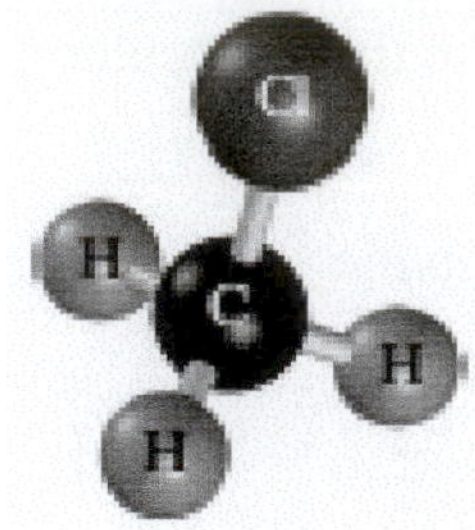

Many organic compounds, especially those of biological interest, contain oxygen as well as carbon and hydrogen. Oxygen has a bond requirement of two, and this may be satisfied in a number of ways. If an –OH is attached to a carbon, an *alcohol* is formed.

alcohol CH_3CH_2-OH

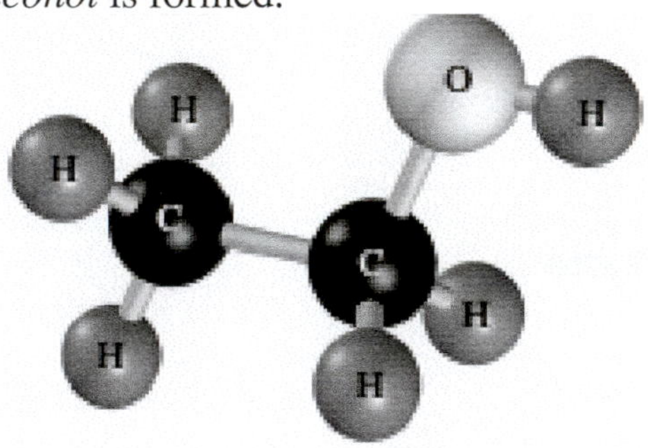

A *phenol* is a special class of alcohol where the –OH is attached directly to a benzene ring.

phenol C_6H_5-OH

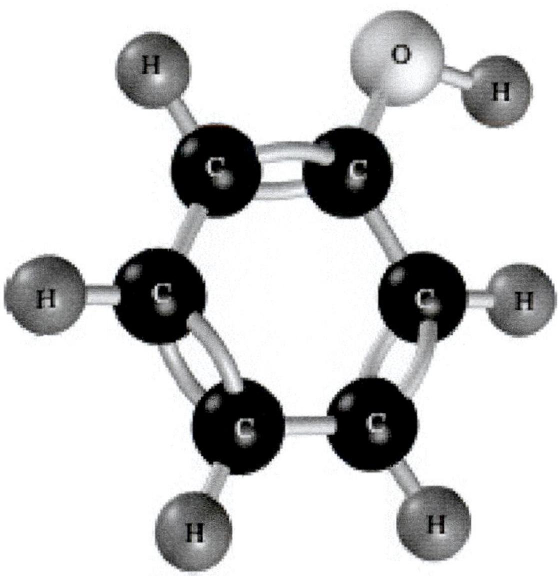

An oxygen atom may also be bonded to two carbon atoms. If that is the case, the class of compound is an *ether*.

ether CH_3-O-CH_3

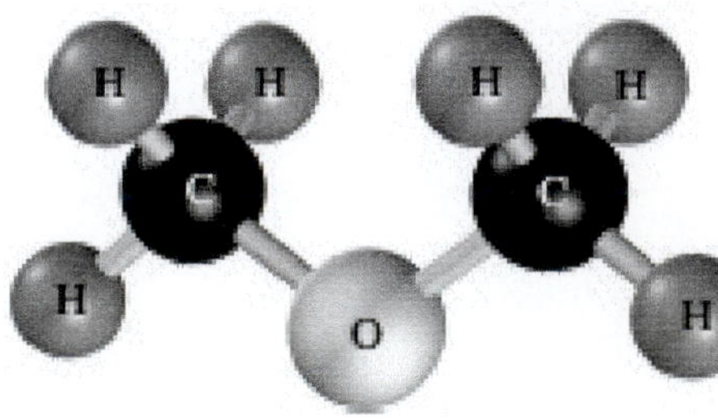

If we attach $-NH_2$ directly to a hydrocarbon, an *amine* results. Low-molecular-weight amines smell like ammonia, but develop a fishy smell as the hydrocarbon chain increases in length.

amine CH_3-NH_2

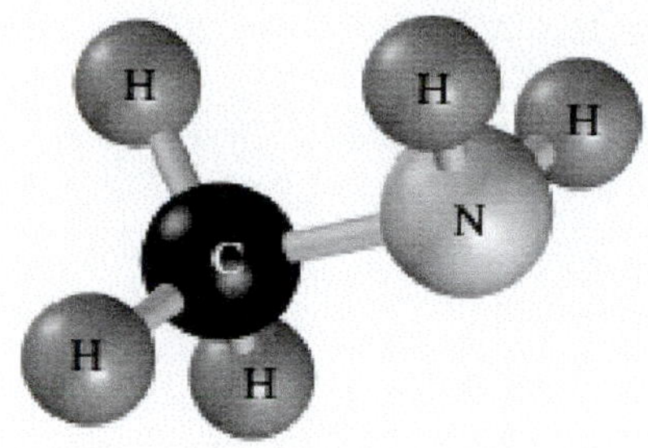

Oxygen can also satisfy its bond requirement of two by forming a double bond with carbon; this is called a **carbonyl group**.

$$-\overset{\overset{\displaystyle O}{\|}}{C}-$$

If the carbonyl group is at the end of the molecule—that is, the carbonyl is bonded to a hydrogen atom—the functional group is an *aldehyde*. If the carbonyl group is in the middle of the carbon chain—that is, the carbonyl is bonded to two other carbon atoms—the functional group is a *ketone*.

aldehyde $CH_3-CH_2-\overset{\overset{\displaystyle O}{\|}}{C}-H$

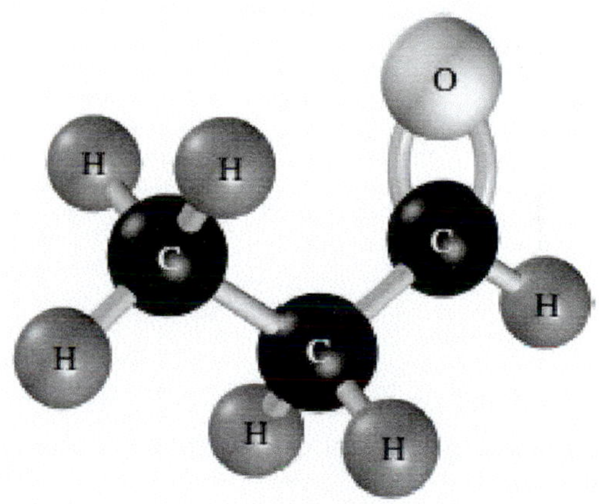

ketone $CH_3-\overset{\overset{\displaystyle O}{\|}}{C}-CH_3$

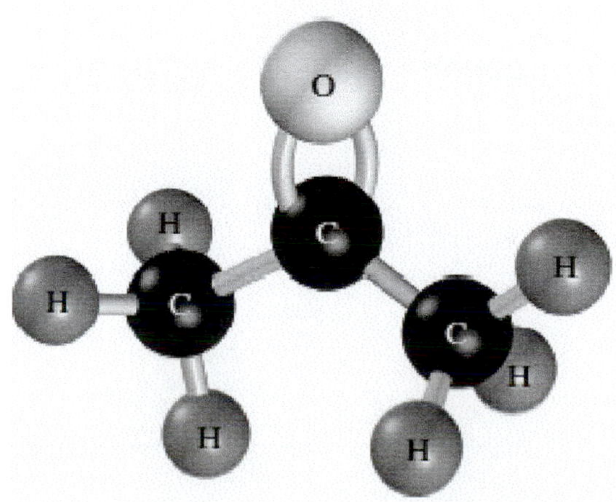

Another possibility is that the carbonyl group may be attached to an –OH group. The resulting structure is found in the class of compounds called *carboxylic acids*.

carboxylic acid $CH_3-\overset{\displaystyle O}{\overset{\|}{C}}-OH$

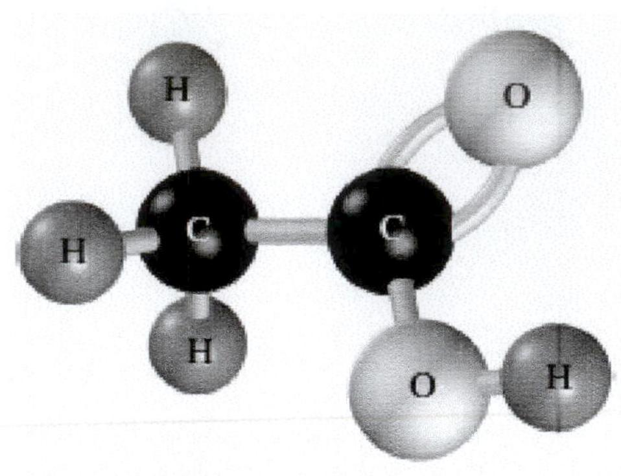

If we substitute a carbon atom for the hydrogen in the carboxylic acid group, we have a class of compounds called *esters*. Esters are noted for their typically fragrant odors.

ester $CH_3-\overset{\displaystyle O}{\overset{\|}{C}}-O-CH_3$

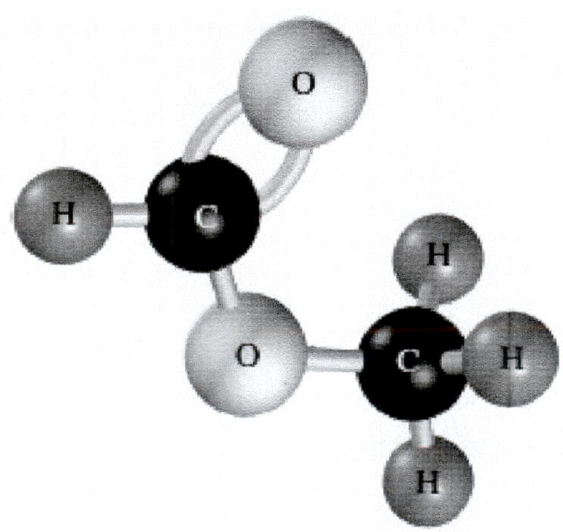

Starting with a carboxylic acid, we can remove the –OH group and replace it with $-NH_2$. The resulting class of compound is called an *amide*.

amide $CH_3-\overset{\displaystyle O}{\overset{\|}{C}}-NH_2$

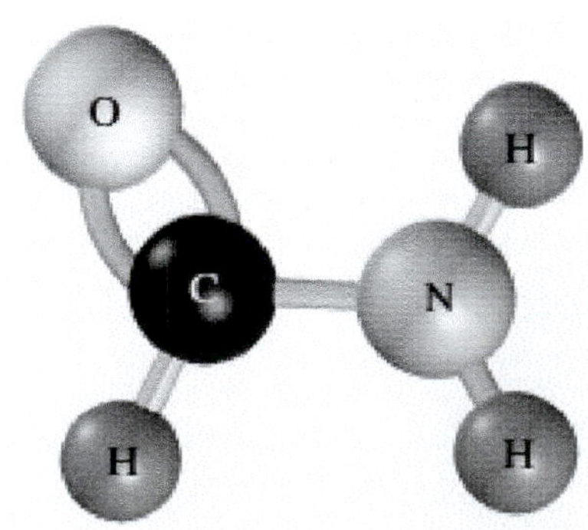

EQUIPMENT and CHEMICALS

- Molecular Model Kit
  Student molecular model sets (ISBN: 0-205-08136-3) are available from Prentice Hall @ 1-800-922-0579 (www.prenhall.com).

Directions for Using Molecular Models

When constructing a model, a rigid connector between two balls represents a single bond. If two balls are joined by two flexible connectors, the two connectors represent a double bond. If two balls are joined by three flexible connectors, the three connectors represent a triple bond.

one rigid connector	—	single bond (one e– pair)
two flexible connectors	—	double bond (two e– pairs)
three flexible connectors	—	triple bond (three e– pairs)

A molecular model uses different color balls to represent hydrogen, carbon, oxygen, chlorine, and nitrogen atoms. The color code for each ball is as follows:

white ball	—	hydrogen atom (one hole)
black ball	—	carbon atom (four holes)
red ball	—	oxygen atom (two holes)
green ball	—	chlorine atom (one hole)
orange ball	—	bromine atom (one hole)
purple ball	—	iodine atom (one hole)
blue ball	—	nitrogen atom (three holes)

Note: If the blue nitrogen ball has more than three holes, use a small peg or tape to fill the additional hole(s). All holes in each ball must have a connector for a model to be built correctly.

PROCEDURE

A. Molecular Models of Hydrocarbons*

Construct the molecular models for the following hydrocarbons and draw their expanded structural formulas in the Data Table.

1. *Alkanes*
   (a) methane, CH_4
   (b) ethane, C_2H_6
   (c) propane, C_3H_8
   (d) butane, C_4H_{10}

2. *Alkenes*
   (a) ethene ("ethylene"), C_2H_4
   (b) propene ("propylene"), C_3H_6

3. *Alkynes*
   (a) ethyne ("acetylene"), C_2H_2
   (b) propyne ("methyl acetylene"), C_3H_4

4. *Arenes*
   (a) methyl benzene, $C_6H_5{-}CH_3$
   (b) *para*-dimethyl benzene, $C_6H_4{-}(CH_3)_2$

* A chemical name in quotation marks indicates a common name rather than a systematic IUPAC name.

B. Molecular Models of Hydrocarbon Derivatives

Construct the molecular models for the following hydrocarbon derivatives, and draw their expanded structural formulas in the Data Table.

1. *Organic Halides*
   (a) "methyl chloride," CH_3–Cl (b) "ethyl iodide," CH_3CH_2–I

2. *Alcohols*
   (a) methanol, CH_3–OH ("methyl alcohol") (b) ethanol, CH_3CH_2–OH ("ethyl alcohol")
   (c) 1–propanol, $CH_3CH_2CH_2$–OH ("propyl alcohol") (d) 2–propanol, $CH_3CH(OH)CH_3$ ("isopropyl alcohol")

3. *Phenols*
   (a) phenol, C_6H_5–OH (b) *ortho*-methyl phenol, $C_6H_4(CH_3)$–OH

4. *Ethers*
   (a) "dimethyl ether," CH_3–O–CH_3 (b) "diethyl ether," CH_3CH_2–O–CH_2CH_3

5. *Amines*
   (a) "methyl amine," CH_3–NH_2 (b) "ethyl amine," CH_3CH_2–NH_2
   (c) "propyl amine," $CH_3CH_2CH_2$–NH_2 (d) "isopropyl amine," $CH_3CH(NH_2)CH_3$

6. *Aldehydes*
   (a) methanal, HCHO ("formaldehyde") (b) ethanal, CH_3CHO ("acetaldehyde")

7. *Ketones*
   (a) propanone, CH_3COCH_3 ("acetone" or "dimethyl ketone") (b) butanone, $CH_3COCH_2CH_3$ ("methyl ethyl ketone")

8. *Carboxylic Acids*
   (a) methanoic acid, HCOOH ("formic acid") (b) ethanoic acid, CH_3COOH ("acetic acid")

9. *Esters*
   (a) methyl methanoate, $HCOOCH_3$ ("methyl formate") (b) ethyl ethanoate, $CH_3COOCH_2CH_3$ ("ethyl acetate")

10. *Amides*
    (a) methanamide, $HCONH_2$ ("formamide") (b) ethanamide, CH_3CONH_2 ("acetamide")

C. Unknown Molecular Models

The Instructor will provide numbered molecular models of unknown organic compounds. Draw the expanded structural formula in the Data Table and identify the class of compound for the following hydrocarbon derivatives: ***organic halide, alcohol, phenol, ether, amine, aldehyde, ketone, carboxylic acid, ester***, and ***amide***.

EXPERIMENT 23 NAME ____________________

DATE ____________ SECTION ____________________

*PRELABORATORY ASSIGNMENT**

1. Provide the key term that corresponds to each of the following definitions.

________ (a) the study of carbon-containing compounds

________ (b) a family of compounds in which all the members have the same structural feature and similar chemical properties

________ (c) an atom or group of atoms that characterizes a class of compounds

________ (d) compounds with the same molecular formula but different structural formulas

________ (e) a compound containing only hydrogen and carbon

________ (f) a hydrocarbon containing only single bonds

________ (g) a hydrocarbon containing a double bond or triple bond

________ (h) a compound containing carbon, hydrogen, and another element such as oxygen, nitrogen, or a halogen

________ (i) the C=O group, which is present in aldehydes, ketones, carboxylic acids, esters, and amides

Key Terms: carbonyl group, class of compounds, functional group, hydrocarbon, hydrocarbon derivative, isomers, organic chemistry, saturated hydrocarbons, unsaturated hydrocarbons

2. What do each of the following represent in the molecular model kit?

(a) one rigid connector (b) two flexible connectors

(c) three flexible connectors (d) white ball

(e) black ball (f) red ball

(g) green ball (h) orange ball

(i) purple ball (j) blue ball

3. Draw the structural formula for a black ball and green ball joined by one rigid connector.

4. Draw the structural formula for a black ball and red ball joined by two flexible connectors.

---

* *Answers in Appendix J*

5. State the name for each of the following alkyl groups.

(a) CH_3- (b) CH_3-CH_2-

(c) $CH_3-CH-CH_3$ (d) $CH_3-CH_2-CH_2-$

6. Identify the three isomers of dichlorobenzene.

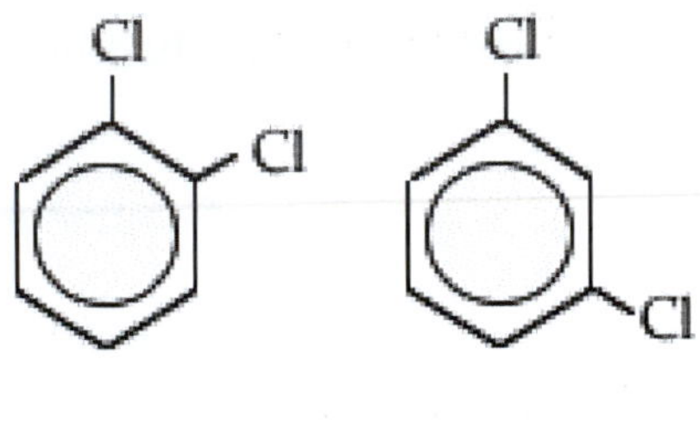

7. Identify the class of compound for each of the following functional groups.

(a) 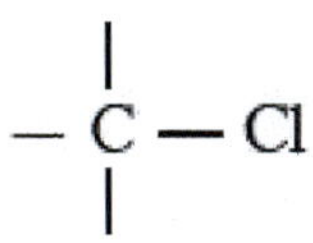

(b) 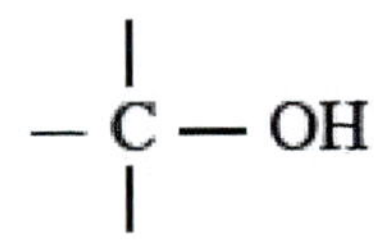

(c) C_6H_5-OH

(d) 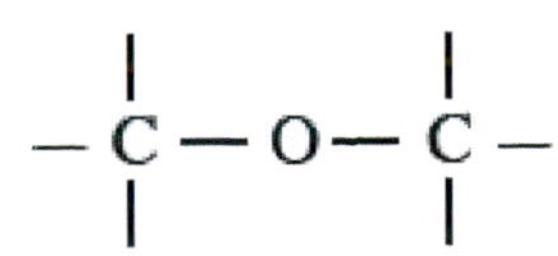

(e) 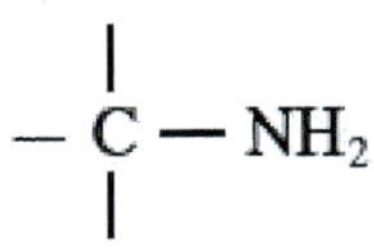

(f) 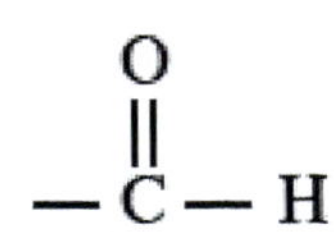

(g) 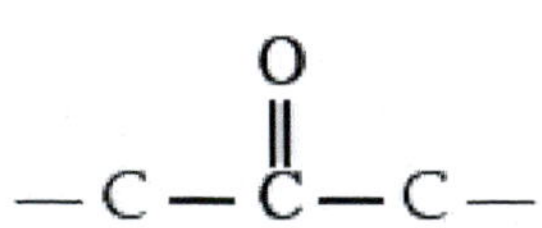

(h) 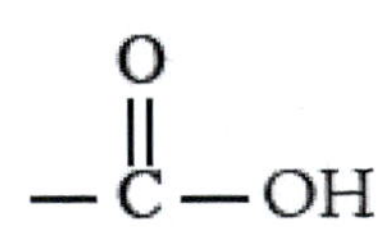

(i) 

(j) 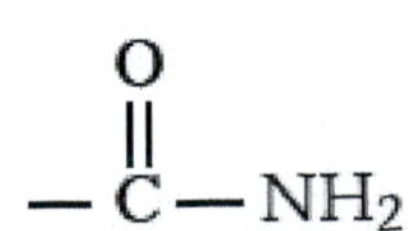

EXPERIMENT 23 NAME ______________________

DATE ______________ SECTION ______________________

DATA TABLE

A. Molecular Models of Hydrocarbons **Model Kit #_______**

1. *Alkanes*
   (a) methane, CH_4
   (b) ethane, C_2H_6
   (c) propane, C_3H_8
   (d) butane, C_4H_{10}

2. *Alkenes*
   (a) ethene ("ethylene"), C_2H_4
   (b) propene ("propylene"), C_3H_6

3. *Alkynes*
   (a) ethyne ("acetylene"), C_2H_2
   (b) propyne ("methyl acetylene"), C_3H_4

4. *Arenes*
   (a) methyl benzene, $C_6H_5–CH_3$
   (b) *para*-dimethyl benzene, $C_6H_4–(CH_3)_2$

B. Molecular Models of Hydrocarbon Derivatives

1. *Organic Halides*
   (a) "methyl chloride," $CH_3–Cl$
   (b) "ethyl iodide," $CH_3CH_2–I$

2. *Alcohols*
   (a) methanol, $CH_3–OH$ ("methyl alcohol")
   (b) ethanol, $CH_3CH_2–OH$ ("ethyl alcohol")
   (c) 1–propanol, $CH_3CH_2CH_2–OH$ ("propyl alcohol")
   (d) 2–propanol, $CH_3CH(OH)CH_3$ ("isopropyl alcohol")

3. *Phenols*
   (a) phenol, $C_6H_5–OH$
   (b) *ortho*-methyl phenol, $C_6H_4(CH_3)–OH$

4. *Ethers*
   (a) "dimethyl ether," $CH_3–O–CH_3$
   (b) "diethyl ether," $CH_3CH_2–O–CH_2CH_3$

5. *Amines*
   (a) "methyl amine," $CH_3–NH_2$
   (b) "ethyl amine," $CH_3CH_2–NH_2$
   (c) "propyl amine," $CH_3CH_2CH_2–NH_2$
   (d) "isopropyl amine," $CH_3CH(NH_2)CH_3$

6. *Aldehydes*
   (a) methanal, HCHO ("formaldehyde")
   (b) ethanal, CH_3CHO ("acetaldehyde")

7. *Ketones*
   (a) propanone, CH_3COCH_3 ("acetone" or "dimethyl ketone")
   (b) butanone, $CH_3COCH_2CH_3$ ("methyl ethyl ketone")

8. *Carboxylic Acids*
   (a) methanoic acid, HCOOH ("formic acid")
   (b) ethanoic acid, CH_3COOH ("acetic acid")

9. *Esters*
   (a) methyl methanoate, $HCOOCH_3$ ("methyl formate")
   (b) ethyl ethanoate, $CH_3COOCH_2CH_3$ ("ethyl acetate")

10. *Amides*
    (a) methanamide, $HCONH_2$ ("formamide")
    (b) ethanamide, CH_3CONH_2 ("acetamide")

C. Unknown Molecular Models

Model Number	*Structural Formula*	*Class of Compound*
#1		#1 ____________________
#2		#2 ____________________
#3		#3 ____________________
#4		#4 ____________________
#5		#5 ____________________
#6		#6 ____________________
#7		#7 ____________________
#8		#8 ____________________
#9		#9 ____________________
#10		#10 ____________________

EXPERIMENT 23 NAME ______________________

DATE ______________ SECTION ______________________

POSTLABORATORY ASSIGNMENT

1. The structural formula for acetylsalicylic acid (aspirin) is shown below. Identify each of the functional groups circled.

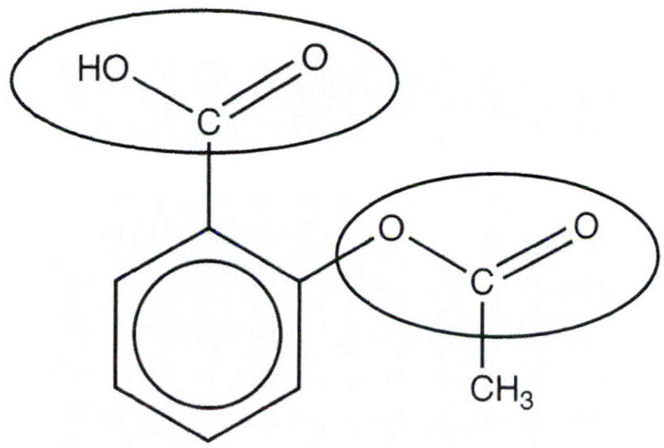

2. The structural formula for cinnamaldehyde (cinnamon flavor) is shown below. Identify each of the functional groups circled.

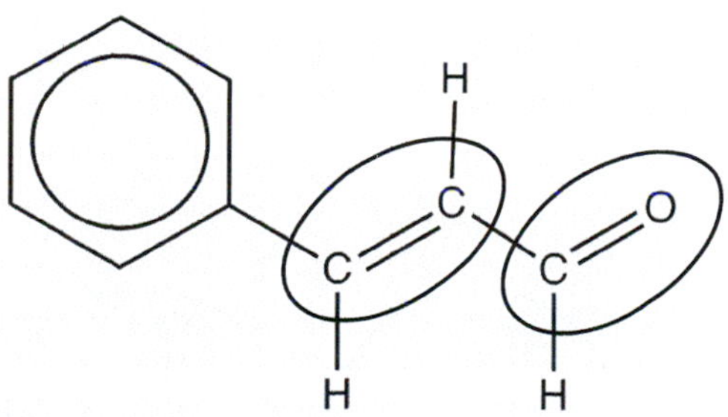

3. The structural formula for chloroacetophenone ("tear gas") is shown below. Identify each of the functional groups circled.

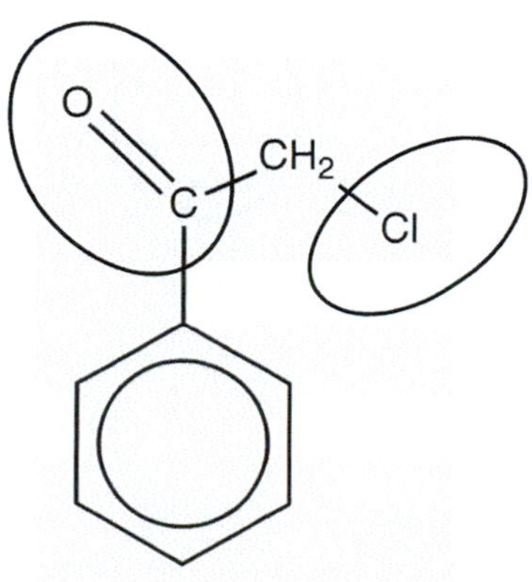

4. Draw two isomers for the molecular formula C_2H_6O. Identify the class of compound for each isomer.

5. Draw two isomers for the molecular formula $C_2H_4O_2$. Identify the class of compound for each isomer.

6. (optional) The structural formula for the artificial sweetener aspartame (NutraSweet®) is shown below. Identify each of the functional groups circled.

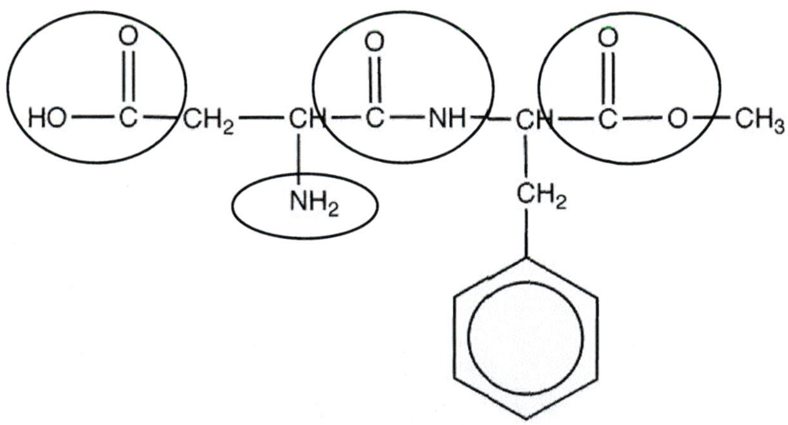

EXPERIMENT

24

Separation of Food Colors and Amino Acids

OBJECTIVES

- To separate blue, green, red, and yellow food colors by paper chromatography.
- To identify an amino acid in an unknown solution by paper chromatography.
- To develop the lab skill for preparing and developing a paper chromatogram.

DISCUSSION

Chromatography (kro-muh-TOG-ruh-fee) is a method for separating a chemical mixture into its component compounds. The term chromatography is derived from the Greek *chroma,* meaning "color" and *graph,* meaning "record." The Russian botanist Mikhail Tsvett is credited with the first application of the technique, when he separated colored plant pigments in the early 1900s.

Chromatography is a powerful tool that requires only tiny amounts of sample and is routinely employed in **biochemistry** analyses. The technique involves putting a small drop of solution on an adsorbent, such as paper or gel, and placing the sample in a chamber with a liquid solvent. In paper chromatography, a drop of sample is placed on a sheet of paper, and the compounds in the mixture separate as the solvent travels up the paper by capillary action.

A compound in a mixture that is more strongly attracted to the solvent travels farther up the paper. A compound less attracted to the solvent travels less, and remains closer to the initial spot. Thus, as the solvent travels up the paper, the compounds in a mixture begin to separate. Food dyes separate into colored, visible spots. However, amino acids separate into colorless, invisible spots, and we must treat the paper to make the spots visible. For example, if we spray the chromatogram with the chemical ninhydrin, amino acid spots change from colorless to purple.

Every compound is characterized by its attraction to the solvent and the adsorbent paper. A retention factor (R_f) indicates the relative attraction of the compound for solvent and for the paper. The **R_f value** expresses the ratio of the distance traveled by the compound compared to the distance traveled by the solvent. The leading edge of the solvent is called the **solvent front**.

Example Exercise 24.1 • Determining an R_f Value

Paper chromatography is used to analyze four biological samples (A, B, C, D), which give the following result using alcohol solvent. State the number of components in each sample, and calculate the R_f value for the compound shown in lane 1.

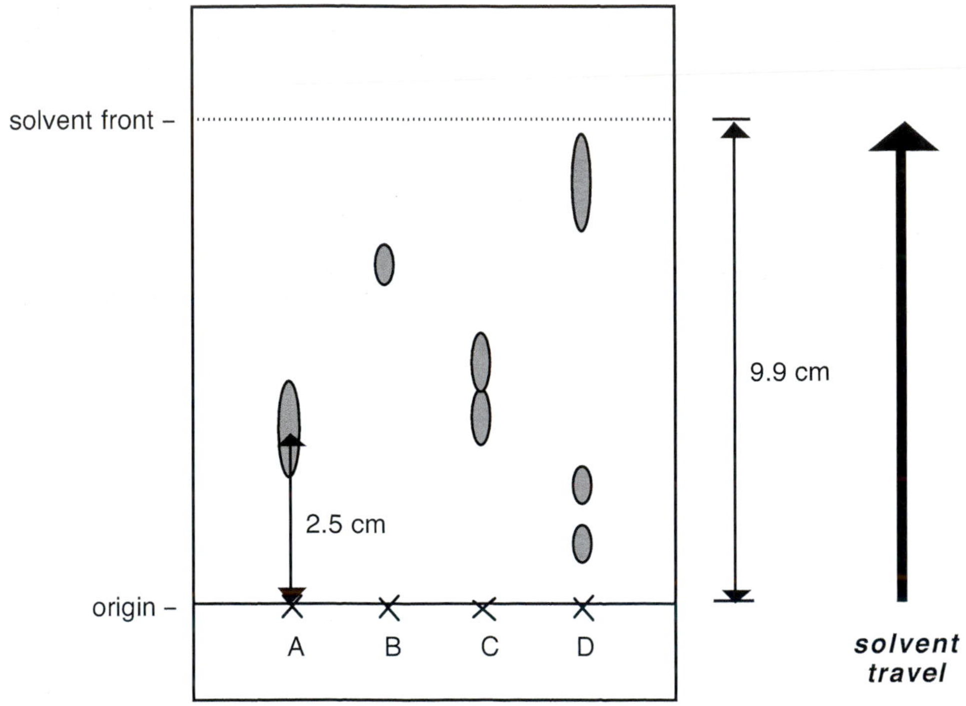

Solution: Samples A and B each have one component as shown by a single spot in lanes 1 and 2. Sample C has two components and D has three components as shown by the spots in lanes 3 and 4.

The distance from the origin to the solvent front is 9.9 cm. The distance from the origin to the center of the spot in lane 1 is 2.5 cm. Therefore, we can calculate the R_f value for the compound in lane 1 as follows.

$$R_f = \frac{2.5\ \cancel{cm}}{9.9\ \cancel{cm}} = 0.25$$

We can describe the theory of paper chromatography as follows. Cellulose paper adsorbs moisture from the air and forms a **stationary phase** as the water vapor is very strongly attracted to the paper. After "spotting" a paper chromatogram, it is placed in a developing chamber containing a liquid solvent. The solvent is a **mobile phase** that travels up the paper and across the spot. As the solvent travels across the sample mixture, it begins pulling compounds up the paper. To understand the process, visualize two liquid sheets moving across one another while competing for a compound. This competition—between the mobile phase and the stationary phase—is responsible for separating the components in a mixture.

EQUIPMENT and CHEMICALS

- Whatman No. 1 chromatography paper
- glass capillary tubes for spotting samples
- 1000-mL beaker
- glass stirring rod
- adhesive tape
- graduated cylinder
- long-stem funnel
- aluminum foil, ~15 x 15 cm square
- latex gloves
- 110 °C drying oven (or heat lamp)
- metric ruler
- blue, green, red, yellow food colors
- solvent for developing chamber (1:1:1 solvent mixture of 2 *M* NH_4OH: pentanol: ethanol)
- known amino acid solutions
  glycine solution, 0.1 *M* Gly
  alanine solution, 0.1 *M* Ala
  phenylalanine solution, 0.1 *M* Phe
- unknown amino acid solutions
  0.1 *M* Gly, 0.1 *M* Ala, 0.1 *M* Phe
- 2% ninhydrin in ethanol

PROCEDURE

A. Separation of Food Colors by Paper Chromatography

1. Cut a piece of chromatography paper that measures 9.0 x 15.0 cm. Using a pencil, draw a line 2.0 cm from the bottom of the paper, and mark four equidistant points (x) along the origin line. Label the four points B, G, R, and Y as shown below.

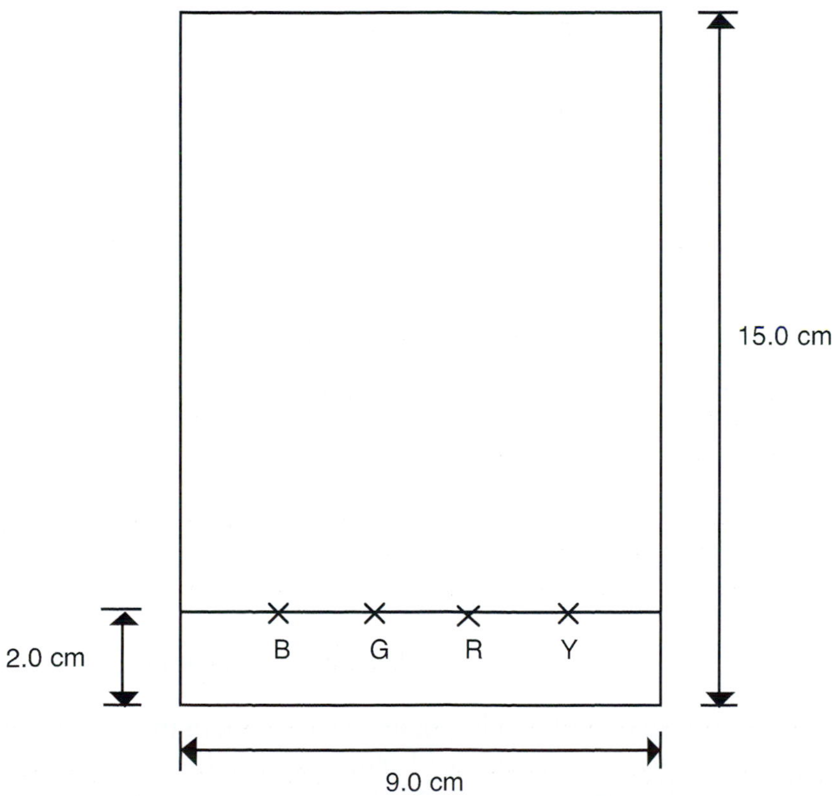

2. Using separate glass capillary tubes, make a 2-mm spot (•) on the origin line at B, G, R, and Y with blue, green, red, and yellow food colors, respectively.

Note: If the initial spot is too large, the colors can bleed into an adjacent spot and confuse the results. To avoid a large initial spot, practice spotting by using the capillary tube on a scrap of chromatography paper.

3. Adjust the length of the chromatogram to fit a 1000-mL beaker, and tape the top edge of the paper to a glass stirring rod as illustrated in Figure 24.1.

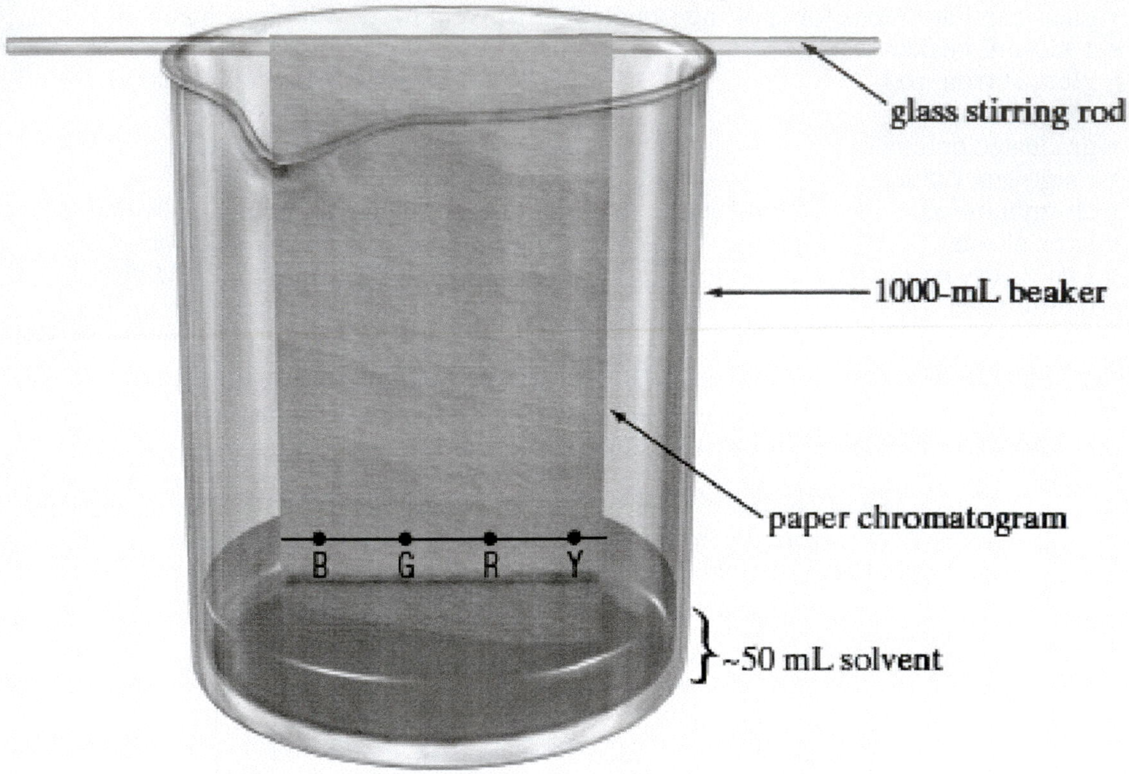

Figure 24.1 Apparatus for Developing a Paper Chromatogram
The initial solvent level must reach the bottom edge of the paper, but should not touch the sample spots on the origin line.

RECYCLE
Chemical
Waste

4. Measure ~50 mL of solvent into a graduated cylinder, and without splashing pour the solvent through a long-stem funnel into the bottom of the 1000-mL beaker. Cover the top of the beaker with a piece of aluminum foil.

   Note: While the food colors are separating in the developing chamber, go on to **Procedure B**, Identification of Amino Acids.

5. After the solvent travels about halfway up the paper chromatogram (~1.5 hours), remove the chromatogram, and draw a dashed line (-----) along the solvent front.

6. Place the chromatogram in a drying oven (or under a heat lamp) for ~5 minutes.

7. When the paper is dry, circle each spot and place a dot in the center. Measure the distance from the origin to the solvent front, and from the origin to the center of each spot for the blue (B); green (G); red (R); and yellow (Y) food colors.

   Note: The components in a food color can vary with brand name, but each food color will usually have more than one component.

8. Calculate the R_f value for each component in the blue (B); green (G); red (R); and yellow (Y) food colors.

B. Identification of Amino Acids by Paper Chromatography

1. Put on latex gloves, and cut a rectangular sheet of chromatography paper that measures 9.0 x 15.0 cm. Draw a line 2.0 cm from the bottom edge of the paper, and mark four equidistant points (x) along the origin. Label the four points Gly, Ala, Phe, and Unk as shown below.

2. Record the unknown number for Unk in the Data Table. Using separate capillary tubes, make a 2-mm spot (•) on the origin line at Gly, Ala, Phe, and Unk with glycine, alanine, phenylalanine, and the unknown amino acid solution.

3. Adjust the length of the chromatogram to fit a 1000-mL beaker, and tape the top edge of the paper to a glass stirring rod as illustrated in Figure 24.1.

4. Measure ~50 mL of solvent into a graduated cylinder, and carefully (without splashing) pour the solvent through a long-stem funnel into the bottom of the 1000-mL beaker. Cover the top of the beaker with a piece of aluminum foil.

RECYCLE **Chemical Waste**

5. After the solvent travels about halfway up the paper chromatogram (~1.5 hours), remove the chromatogram, and draw a dashed line (-----) along the solvent front.

6. Place the chromatogram in a drying oven (or under a heat lamp) for ~5 minutes.

7. When the paper is dry, spray the chromatogram with ninhydrin solution under a hood (avoid breathing and contact). Dry the chromatogram.

8. Circle each spot on the dry paper with a pencil and place a dot in the center of the spot. Measure the distance from the origin to the solvent front, and from the origin to the center of each spot for Gly, Ala, Phe, and Unk.

9. Calculate the R_f value for glycine (Gly); alanine (Ala); phenylalanine (Phe); and the unknown amino acid (Unk).

10. Refer to R_f values for the known amino acids, and identify the unknown amino acid(s) as Gly, Ala, or Phe.

EXPERIMENT 24 NAME ____________________

DATE ____________________ SECTION ____________________

*PRELABORATORY ASSIGNMENT**

1. Provide the key term that corresponds to each of the following definitions.

__________ (a) the study of compounds derived from plants and animals

__________ (b) a method for separating a mixture into its components as a result of a varying attraction of compounds for a mobile solvent on a stationary solid

__________ (c) the leading edge of the solvent, which travels from the bottom of the developing chamber to the upper portion of the chromatogram

__________ (d) refers to solvent that travels up a paper chromatogram by capillary action

__________ (e) refers to the moisture that is strongly adsorbed onto a paper chromatogram and is not free to travel

__________ (f) the ratio of the distance traveled by a sample component compared to the distance traveled by the solvent

Key Terms: biochemistry, chromatography, mobile phase, R_f value, solvent front, stationary phase

2. Where is the "origin" on a chromatogram?

3. Where is the "solvent front" on a chromatogram?

4. What is meant by "spotting" a chromatogram?

5. Where is "lane 1" on a chromatogram?

Where is the "first" component in a given lane?

---

* *Answers in Appendix J*

6. Why is it necessary to wear latex gloves when handling the amino acid chromatogram?

7. An amino acid travels 2.5 cm from the origin on a paper chromatogram, and the solvent front travels 7.5 cm. Show the calculation for the R_f value of the amino acid?

8. Why should a pencil—*not a pen*—be used to mark the origin on a paper chromatogram?

9. What precautions should be observed in this experiment?

EXPERIMENT 24 NAME ____________________

DATE ____________ SECTION ____________________

DATA TABLE

A. Separation of Food Colors by Paper Chromatography

Attach the food color paper chromatogram.

RECYCLE
Chemical
Waste

distance from origin to solvent front ______ cm

distance from origin	**Blue** (B)	**Green** (G)	**Red** (R)	**Yellow** (Y)
to 1st component	______ cm	______ cm	______ cm	______ cm
to 2nd component	______ cm	______ cm	______ cm	______ cm

Calculate the R_f value for each component in the food colors (see Example Exercise 24.1).

1st component	______	______	______	______
2nd component	______	______	______	______

B. Identification of Amino Acids by Paper Chromatography **UNKNOWN #** _______

Circle Amino Acid(s) in the Unknown: **Gly** **Ala** **Phe**

Attach the amino acid paper chromatogram.

RECYCLE
Chemical
Waste

distance from origin to solvent front _______ cm

distance from origin	**(Gly)**	**(Ala)**	**(Phe)**	**(Unk)**
to amino acid	_______ cm	_______ cm	_______ cm	_______ cm

Calculate the R_f value for each amino acid (see Example Exercise 24.1).

amino acid	_______	_______	_______	_______

Identify the amino acid(s) in the unknown (Unk) from the R_f value.

____

EXPERIMENT 24 NAME ______________________________

DATE ____________________ SECTION ______________________________

POSTLABORATORY ASSIGNMENT

1. NutraSweet® is a dipeptide of two amino acids that are released when the artificial sweetener is treated with dilute acid. Following treatment with acid, NutraSweet® (NuS) was spotted on a paper chromatogram along with amino acid solutions of alanine (Ala), aspartic acid (Asp), glycine (Gly), lysine (Lys), phenylalanine (Phe), tryptophan (Trp), and valine (Val).

   The paper chromatogram was placed in a solvent chamber with the following results.

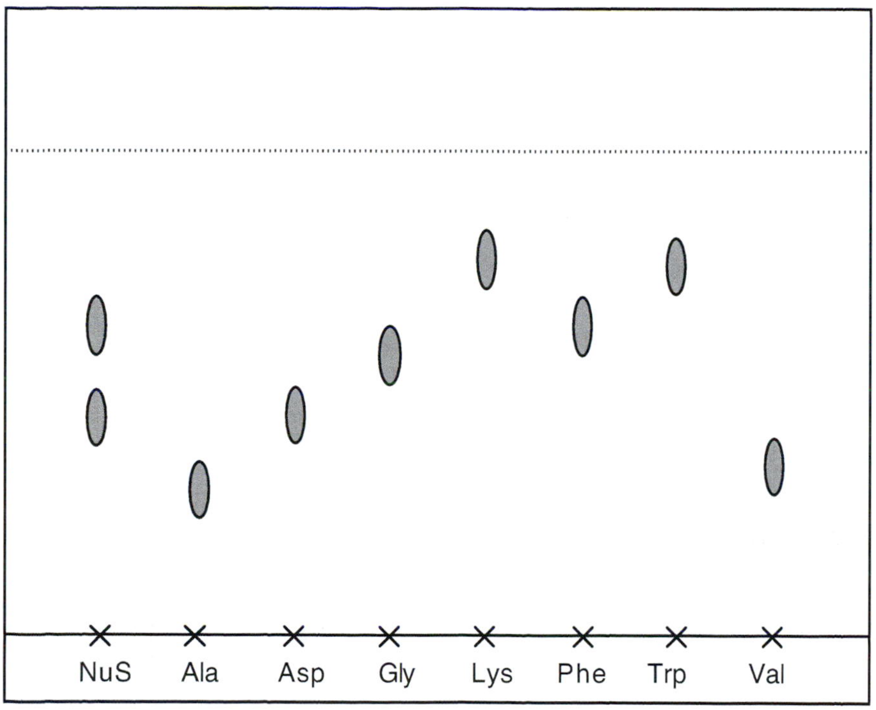

   (a) What are the two amino acids in a molecule of NutraSweet®? Explain. ______

   (b) What is the term for the amide bond in a protein that joins two amino acids? ______

   (c) What is the term for a protein molecule that catalyzes (speeds up) a biochemical reaction? ______

2. The genetic code specifies how ribonucleic acid (RNA) synthesizes proteins by sequentially joining amino acids together. A segment of RNA contains codons composed of three of the four bases: adenine (A), cytosine (C), guanine (G), or uracil (U).

   A GGU codon specifies the addition of glycine to a growing protein chain; a GCU codon specifies the addition of alanine, and a UUU codon specifies phenylalanine. A "stop" codon terminates the addition of amino acids to a growing protein chain.

   A solution of the "stop" codon, XXX, was spotted on a paper chromatogram along with the codons AAA, AUG, CGC, CUG, GAG, UAA, and UUU. The paper chromatogram was placed in a solvent chamber and gave the following results.

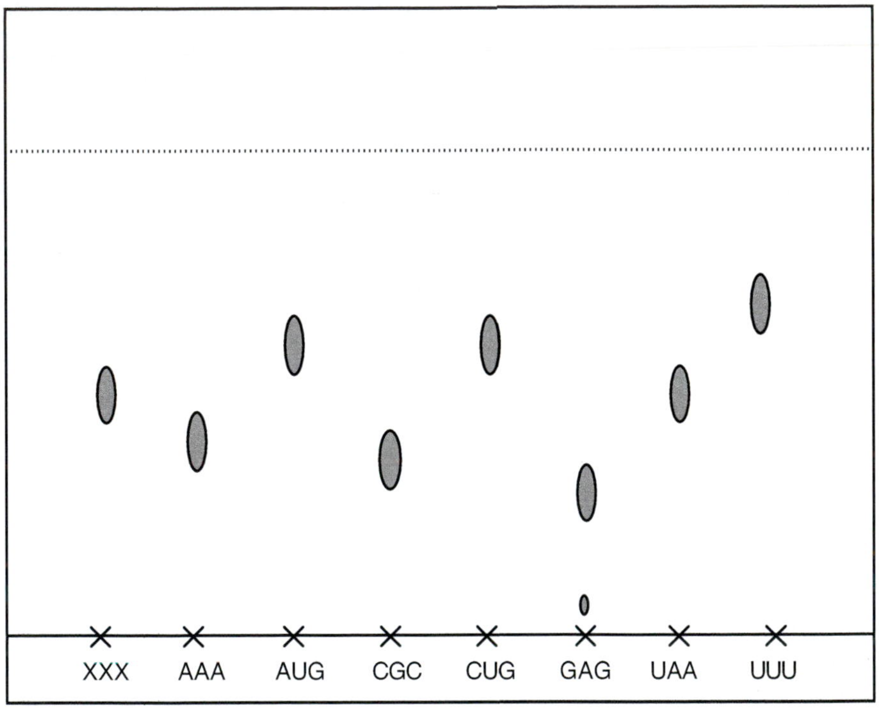

   (a) Based on the chromatogram, what is the identity of the "stop" codon? Explain. ______

   (b) Based on the chromatogram, are any of the solutions contaminated? Explain. ______

3. (optional) A nucleotide is a repeating unit in a nucleic acid. A nucleotide is composed of a sugar, a base, and phosphoric acid. State the nucleic acid (DNA or RNA) containing:

   (a) ribose sugar, uracil base, and phosphoric acid ______

   (b) deoxyribose sugar, thymine base, and phosphoric acid ______

EXPERIMENT

25

Laboratory Instruments and Techniques

OBJECTIVES

- To correctly record measurements for length, mass, volume, and temperature.
- To state the uncertainty in a measurement obtained from a metric ruler, balance, graduated cylinder, pipet, buret, and thermometer.
- To round off nonsignificant digits correctly when adding, subtracting, multiplying, and dividing laboratory data.
- To demonstrate skill in performing laboratory procedures.
- To state the precautions regarding laboratory safety.

DISCUSSION

One prominent educational theory states that learning science occurs in three domains; that is, cognitive, affective, and motor skill. In the cognitive domain, students develop intellectual skills such as problem solving and concept modeling. In the affective domain, students gain an appreciation for the value of science in society and begin to understand both the benefits and limitations of science. In the motor skill domain, students acquire those skills that scientists employ in the laboratory. These skills include the proper handling of chemicals and glassware, as well as the accurate reading of scientific instruments.

Introductory chemistry experiments provide a unique opportunity to develop "hands-on" skills in the responsible handling of chemicals and glassware. In addition, students learn that all scientific measurements have *uncertainty* that is inherent in the instrument that is used for the measurement.

In this experiment, you will demonstrate your ability to take measurements with several instruments that you have previously used in the laboratory. Table 25.1 indicates the uncertainty for each of the instruments in this experiment. Moreover, example readings are provided that are similar to actual measurements taken during the experiment.

Table 25.1 Measurement of Length, Mass, Volume, and Temperature

QUANTITY	*INSTRUMENT*	*UNCERTAINTY*	*EXAMPLE*
length	METRIC RULER A	± 0.1 cm	5.2 cm
	METRIC RULER B	± 0.05 cm	5.25 cm
mass	decigram balance	± 0.1 g	86.3 g
	centigram balance	± 0.01 g	86.32 g
	milligram balance	± 0.001 g	86.318 g
volume	100-mL graduated cylinder	± 0.5 mL	92.5 mL
	10-mL volumetric pipet	± 0.1 mL	10.0 mL
	50-mL buret	± 0.05 mL	36.50 mL
temperature	110 °C thermometer	± 0.5°C	20.5°C

EQUIPMENT and CHEMICALS

- practical exam at numbered stations
- written exam
- safety goggles
- scientific calculator

PROCEDURE

Students are to be seated at numbered stations arranged consecutively about the laboratory. Each station requires the student to answer a brief question regarding an experimental setup or procedure. The student is allowed approximately one minute at each station, at which time the instructor signals students to move to the next station.

NOTE: Record your answers for the *lab practical exam* in the DATA TABLE.
Record your answers for the *lab written exam* directly on the test.

A. Lab Practical Exam

Station #1. Identify laboratory equipment.

Station #2. State the location for chemical waste disposal.

Station #3. Estimate a length in centimeters.

Station #4. Estimate a mass in grams.

Station #5. Estimate a volume in milliliters.

Station #6. Read a metric ruler with centimeter divisions.

Station #7. Read a metric ruler with millimeter subdivisions.

Station #8. Read a decigram balance.

Station #9. Read a centigram balance.

Station #10. Read a milligram balance.

Station #11. Read a graduated cylinder.

Station #12. Read a Celsius thermometer.

Station #13. Classify a sample element as a *metal, semimetal,* or *nonmetal.*

Station #14. Identify a wavelength of light using a hand spectroscope.

Station #15. Identify an *alkali* or *alkaline earth* element from a flame test.

Station #16. Identify *bromide, chloride, or iodide* by observing a halide test.

Station #17. Critique an apparatus for determining a boiling point.

Station #18. Critique an apparatus for decomposing a hydrate.

Station #19. Critique an apparatus for centrifuging a precipitate.

Station #20. Identify a solution as *acidic* or *basic* using litmus paper.

Station #21. Identify the *aqueous layer* in a beaker with two immiscible liquids.

Station #22. Draw the *structural formula* given a molecular model.

Station #23. Draw the *electron dot formula* given a molecular model.

Station #24. Identify the correct technique for using a volumetric pipet.

Station #25. Read a buret.

B. Lab Written Exam

- Identify safety precautions in the laboratory.
- Identify the location for broken glass disposal.
- Identify the instrument for a given measurement.
- Identify the uncertainty for a given instrument.
- Add or subtract experimental data and round off the answer.
- Multiply or divide experimental data and round off the answer.
- Find the mass of a sample given *weighing by difference* data.
- Find the volume of a sample given *volume by displacement* data.
- Distinguish between a *physical property* and a *chemical property*.
- Distinguish between a *physical change* and a *chemical change*.
- Distinguish between a *heterogeneous mixture* and a *homogeneous mixture*.
- Identify the origin of an "atomic fingerprint."
- Identify the *alkali metals*, *alkaline earth metals*, and *halides*.
- Identify a *cation* or *anion* given its chemical behavior.
- Identify five types of chemical reactions.
- Identify four observations that indicate a chemical reaction.
- Distinguish between a *supernate* and a *precipitate*.
- Distinguish between an *empirical formula* and a *molecular formula*.
- Identify the *water of hydration* for a given hydrate.
- Distinguish among *actual yield*, *theoretical yield*, and *percent yield*.
- Distinguish between a "*wet gas*" and a "*dry gas*."
- Identify the *electron pair geometry* for a small molecule.
- Identify the *molecular shape* for a small molecule.
- Estimate the mass of liquid delivered from a volumetric pipet.
- Identify the sources of error in an acid-base titration.

EXPERIMENT 25 NAME ______________________

DATE ______________ SECTION ______________________

*PRELABORATORY ASSIGNMENT**

1. Provide the key term that corresponds to each of the following definitions.

__________ (a) a procedure for obtaining the mass of a sample *indirectly* by first weighing a container and then weighing the container with the sample

__________ (b) a procedure for obtaining the mass of a sample *directly* by first placing a container on an electronic balance and setting the balance to zero; second, add a sample to the container and record the mass of sample directly

__________ (c) to determine the volume of a sample by measuring volume of water displaced

__________ (d) the process of heating, cooling, and weighing until the mass readings for an object are constant, or agree closely

__________ (e) to heat a precipitate in aqueous solution in order to develop larger particles that are easier to filter and free of impurities

__________ (f) refers to liquids that do not dissolve in one another and separate into two layers

__________ (g) to rinse a pipet or buret with a sample to prevent dilution by residual water

__________ (h) the clear lens at the surface of a liquid inside a piece of narrow glassware, such as a graduated cylinder, pipet, or buret

__________ (i) the stage in a titration when the indicator changes color

__________ (j) a procedure for establishing the concentration of a solution precisely

Key Terms: condition, digestion, end point, heating to constant weight, immiscible, meniscus, standardization, tare weighing, volume by displacement, weighing by difference

2. Review each of the following and supply the laboratory manual page reference; e.g., pp. 1-2.

(a) **Safety Precautions** pp. ___

(b) **Locker Inventory** pp. ___

(c) **Chemical Waste Disposal** p. ___

(d) METRIC RULER A shown in **Example Exercise 2.1** p. ___

(e) METRIC RULER B shown in **Example Exercise 2.2** p. ___

(f) GRADUATED CYLINDER shown in **Figure 2.1** p. ___

* *Answers in Appendix J*

(g) THERMOMETER shown in **Figure 2.2** p. ___

(h) BOILING WATER APPARATUS shown in **Figure 2.3** p. ___

(i) WEIGHING BY DIFFERENCE shown in **Figure 3.1** p. ___

(j) VOLUME BY DISPLACEMENT shown in **Figure 3.2** p. ___

(k) MELTING POINT APPARATUS shown in **Figure 4.3** p. ___

(l) CLASSIFICATION OF MATTER shown in **Figure 5.1** p. ___

(m) BOILING POINT APPARATUS shown in **Figure 5.2** p. ___

(n) HAND SPECTROSCOPE shown in **Figures 6.5** p. ____

(o) FLAME-TEST TECHNIQUE shown in **Figures 7.1** and **8.2** pp. ___

(p) LITMUS PAPER TESTS shown in **Figures 8.3** and **9.2** pp. ___

(q) FIRING A CRUCIBLE APPARATUS shown in **Figure 12.1** p. ___

(r) DECOMPOSING A HYDRATE APPARATUS shown in **Figure 13.1** p. ___

(s) GAS COLLECTION APPARATUS shown in **Figures 14.1** p. ___

(t) FILTRATION APPARATUS shown in **Figure 15.1** p. ___

(u) GAS COLLECTION APPARATUS shown in **Figure 16.2** p. ___

(v) GAS COLLECTION APPARATUS shown in **Figure 17.1** p. ___

(w) MOLECULAR MODELS shown in **Experiment 18** p. ___

(x) EVAPORATION APPARATUS shown in **Figure 19.5** p. ___

(y) BURET shown in **Figure 20.1** p. ___

(z) LABORATORY BURNER shown in **Appendix A** p. ___

(aa) DECIGRAM BALANCE shown in **Appendix B** p. ___

(bb) CENTIGRAM BALANCE shown in **Appendix C** p. ___

(cc) MILLIGRAM BALANCE shown in **Appendix D** p. ___

(dd) VOLUMETRIC PIPET shown in **Appendix E** p. ___

3. What precautions should be observed in this experiment?

EXPERIMENT 25 NAME ______________________

DATE ______________ SECTION ______________________

DATA TABLE

A. Lab Practical Exam

Station #1. ______________________

Station #2. ______________________

Station #3. ______________________

Station #4. ______________________

Station #5. ______________________

Station #6. ______________________

Station #7. ______________________

Station #8. ______________________

Station #9. ______________________

Station #10. ______________________

Station #11. ______________________

Station #12. ______________________

Station #13.

Station #14.

Station #15.

Station #16.

Station #17.

Station #18.

Station #19.

Station #20.

Station #21.

Station #22.

Station #23.

Station #24.

Station #25.

B. Lab Written Exam

NOTE: Answer questions on the *written exam* given by the Instructor.

EXPERIMENT 25 NAME ______________________

DATE ______________ SECTION ______________________

POSTLABORATORY ASSIGNMENT

1. What is the name of the porcelain lab equipment shown?

(a)

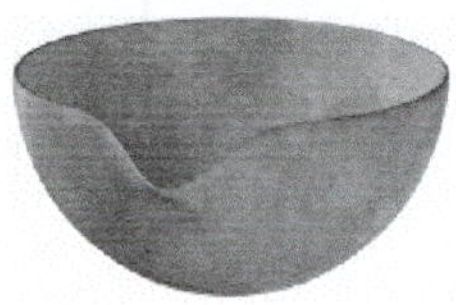

(b)

2. State the uncertainty in measurement for each of the following instruments.

(a) METRIC RULER A ________ (b) METRIC RULER B ________

(c) decigram balance ________ (d) milligram balance ________

(e) graduated cylinder ________ (f) 10–mL pipet ________

(g) 50–mL buret ________ (h) thermometer ________

3. What is the length of the object shown on the metric ruler?

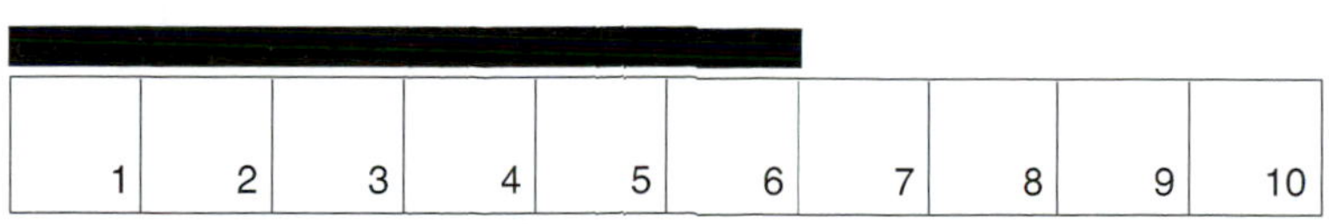

4. What is the length of the object shown on the metric ruler?

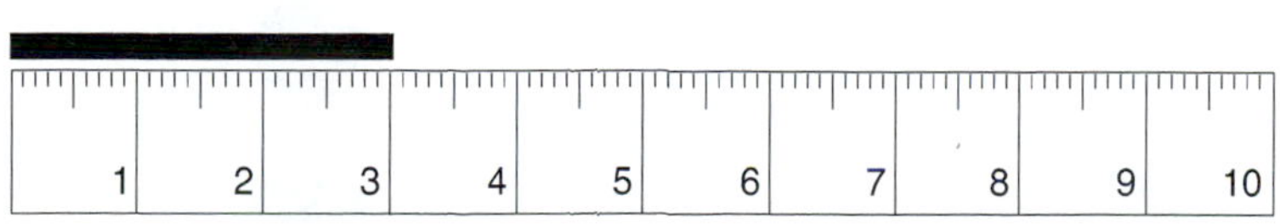

5. What is the volume of liquid shown in the graduated cylinder?

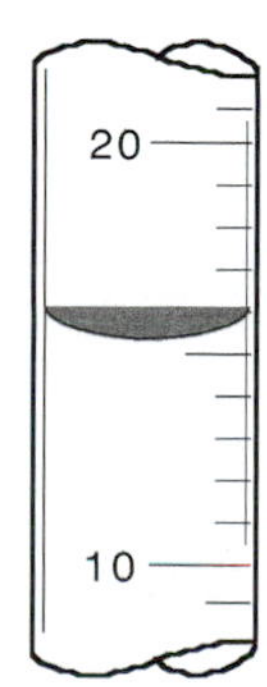

6. What is the temperature indicated by the Celsius thermometer?

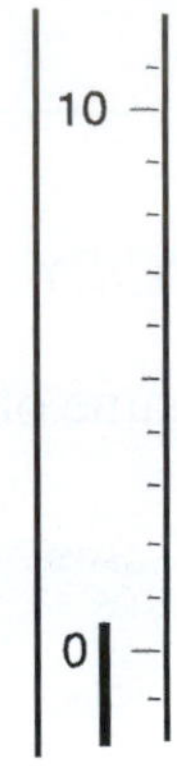

7. What is the volume of solution delivered by a buret as indicated by the initial meniscus and final meniscus shown?

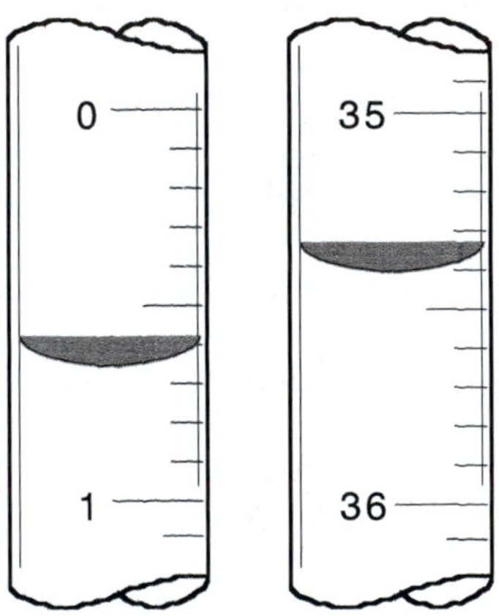

8. (optional) Are the following organic molecular models identical or different? Explain.

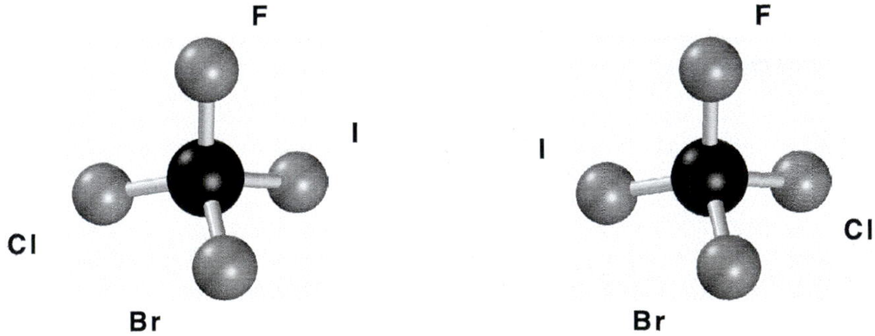

APPENDIX

Appendices

A *Laboratory Burner*

B *Decigram Balance*

C *Centigram Balance*

D *Milligram Balance*

E *Volumetric Pipet*

F *Activity Series for Metals*

G *Solubility Rules*

H *Laboratory Notebook*

I *Glossary*

J *Answers to Prelaboratory Assignments*

APPENDIX

A

Laboratory Burner

Although a variety of lab burners are found in chemistry laboratories, they all employ the same principle. Natural gas is allowed to flow into the barrel of the burner and mix with the air. The ratio of gas to air can be adjusted, which in turn regulates the temperature of the flame. The more air that is available, the hotter the flame. The hottest part of the flame is the tip of the inner pale blue cone. Two typical burners are shown in Figure A.1.

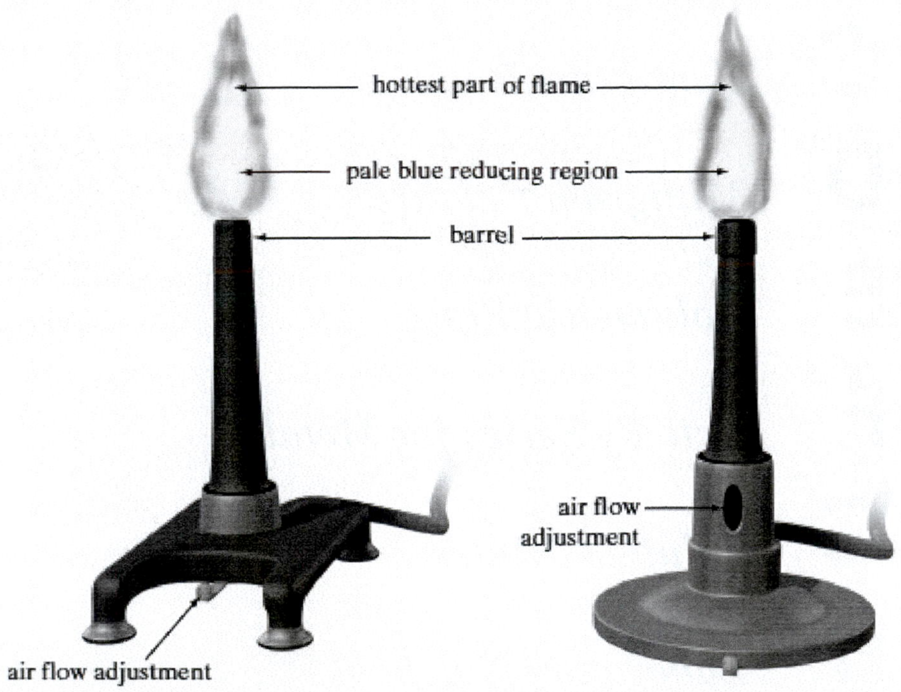

Figure A.1 Laboratory Burners

Steps in Operating a Burner

1. Close the air flow adjustment.
2. Open the gas jet.
3. Light the burner at the top of the barrel.
4. To obtain a hotter flame, open the air flow adjustment.
5. To shut off the burner, close the gas jet.

APPENDIX

B

Decigram Balance

A decigram balance provides measurements with an uncertainty of one decigram (± 0.1 g). The mass is determined by placing a sample on the pan, and adjusting the metal riders on the beams. The heaviest rider (**100-g**) is adjusted first, and the **10-g** rider second. The **1-g** rider is adjusted last and is read to the nearest subdivision (0.1-g). The mass of the sample is equal to the sum of the masses indicated by all of the riders; for example, 100 g + 20 g + 7.2 g = 127.2 g.

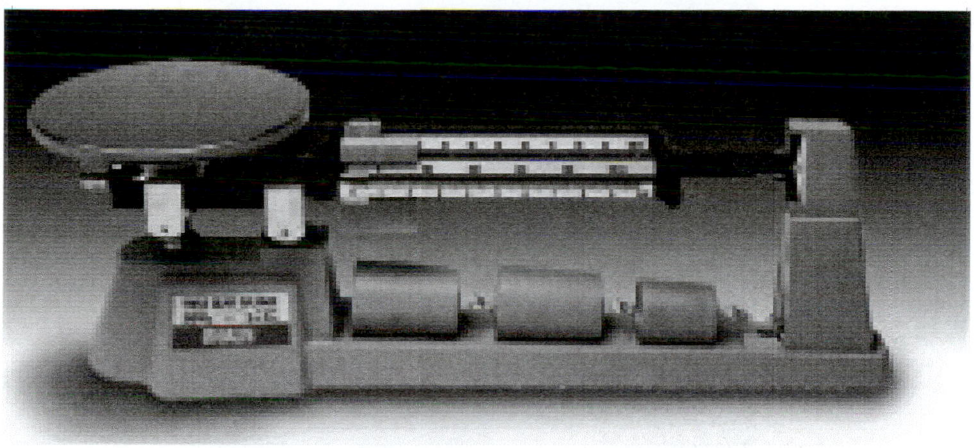

Figure B.1 A beam balance with decigram precision (± 0.1 g). *(Photo courtesy of Ohaus Scale Corporation)*

APPENDIX

C

Centigram Balance

A centigram balance provides measurements with an uncertainty of one centigram (± 0.01 g). The mass is determined by placing a sample on the pan, and adjusting the metal riders on the beams. The heaviest rider (**100-g**) is adjusted first, the **10-g** rider second, and the **1-g** rider third. The **0.1-g** rider is adjusted last and is read to the nearest subdivision (0.01-g). The mass of the sample is equal to the sum of the masses; for example, 100 g + 20 g + 7 g + 0.25 g = 127.25 g.

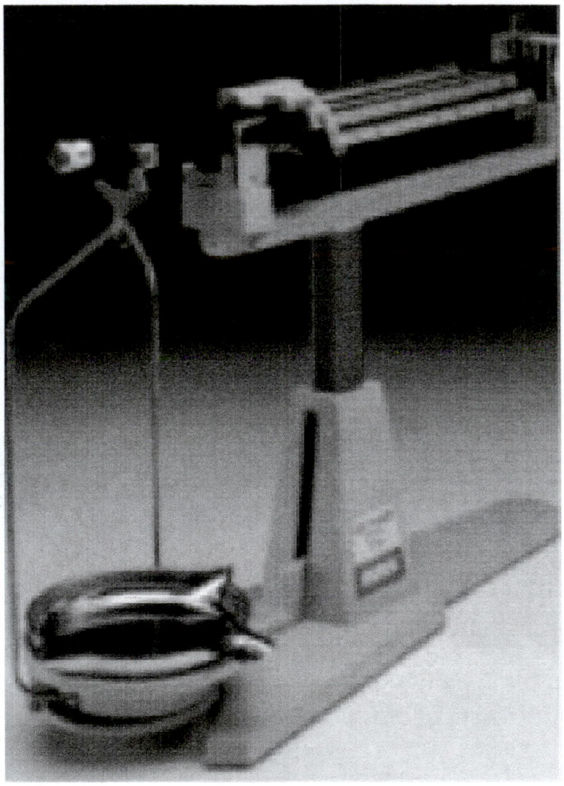

Figure C.1 A beam balance with centigram precision (± 0.01 g). *(Photo courtesy of Ohaus Scale Corporation)*

Milligram Balance

APPENDIX

D

An electronic milligram balance provides measurements with an uncertainty of one milligram (± 0.001 g). The mass is determined by placing a sample on the pan, closing the draft shield doors, and reading the digital display. Tenth milligram electronic balances are also common, which have an uncertainty of one-tenth milligram (± 0.0001 g).

Figure D.1 A digital electronic balance with milligram precision (± 0.001 g). *(Photo courtesy of Mettler Toledo Corporation)*

APPENDIX

E

Volumetric Pipet

When using a **volumetric pipet**, follow these three steps.

1. Condition the pipet with a small portion of the solution to be transferred.
2. Use a pipet bulb to draw the solution above the calibration line as shown in Figure E.1. Slip the pipet bulb off by placing your finger over the end of the pipet. Move your finger slightly so as to allow the bottom of the **meniscus** to drop slowly to the calibration line.
3. Place the tip of the pipet into a flask or beaker. Allow the pipet to drain free and touch off the last drop on the pipet tip (do not blow out the last drop of solution).

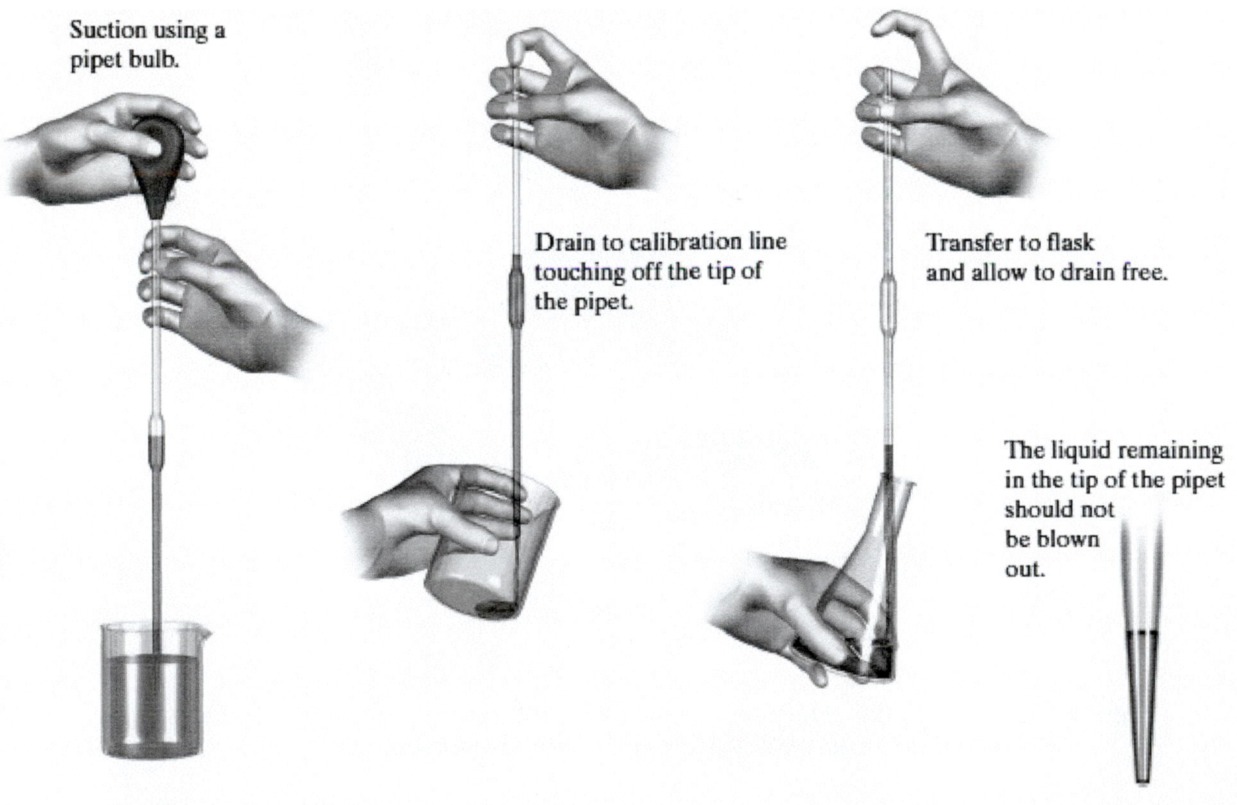

Figure E.1 Transferring a quantity of solution using a volumetric pipet.

APPENDIX

F

Activity Series for Metals

MOST ACTIVE METAL:	$Li \rightarrow Li^{+} + e^{-}$
	$K \rightarrow K^{+} + e^{-}$
	$Ba \rightarrow Ba^{2+} + 2\,e^{-}$
	$Sr \rightarrow Sr^{2+} + 2\,e^{-}$
	$Ca \rightarrow Ca^{2+} + 2\,e^{-}$
	$Na \rightarrow Na^{+} + e^{-}$
(metals above Mg react with water at 25 °C)	
	$Mg \rightarrow Mg^{2+} + 2\,e^{-}$
	$Al \rightarrow Al^{3+} + 3\,e^{-}$
	$Mn \rightarrow Mn^{2+} + 2\,e^{-}$
	$Zn \rightarrow Zn^{2+} + 2\,e^{-}$
	$Cr \rightarrow Cr^{3+} + 3\,e^{-}$
	$Fe \rightarrow Fe^{2+} + 2\,e^{-}$
	$Cd \rightarrow Cd^{2+} + 2\,e^{-}$
	$Co \rightarrow Co^{2+} + 2\,e^{-}$
	$Ni \rightarrow Ni^{2+} + 2\,e^{-}$
	$Sn \rightarrow Sn^{2+} + 2\,e^{-}$
	$Pb \rightarrow Pb^{2+} + 2\,e^{-}$
(metals above H^{+} react with acid)	
	$\mathbf{H_2 \rightarrow 2\,H^{+} + 2\,e^{-}}$
(metals below H_2 do not react with acid)	
	$Cu \rightarrow Cu^{2+} + 2\,e^{-}$
	$Ag \rightarrow Ag^{+} + e^{-}$
	$Hg \rightarrow Hg^{2+} + 2\,e^{-}$
LEAST ACTIVE METAL:	$Au \rightarrow Au^{3+} + 3\,e^{-}$

APPENDIX

G

Solubility Rules

Ionic compounds containing the following ions are generally soluble in water:

1. alkali metal ions and ammonium ions, Li^+, Na^+, K^+, NH_4^+
2. acetate ion, $C_2H_3O_2^-$
3. nitrate ion, NO_3^-
4. halide ions ($X = Cl^-, Br^-, I^-$)
   (AgX, Hg_2X_2, and PbX_2 are exceptions and *insoluble*)
5. sulfate ion, SO_4^{2-}
   ($SrSO_4$, $BaSO_4$, and $PbSO_4$ are exceptions and *insoluble*)

Ionic compounds containing the following ions are generally insoluble* in water:

6. carbonate ion, CO_3^{2-}
   (see Rule 1 exceptions, which are *soluble*)
7. chromate ion, CrO_4^{2-}
   (see Rule 1 exceptions, which are *soluble*)
8. phosphate ion, PO_4^{3-}
   (see Rule 1 exceptions, which are *soluble*)
9. sulfide ion, S^{2-}
   (CaS, SrS, BaS, and Rule 1 exceptions are *soluble*)
10. hydroxide ion, OH^-
   [$Ca(OH)_2$, $Sr(OH)_2$, $Ba(OH)_2$, and Rule 1 are exceptions and *soluble*]

* These ionic compounds are actually slightly soluble, or very slightly soluble, in water.

APPENDIX

H

Laboratory Notebook

The integrity of experimental data is a cornerstone of scientific research. Thus, research data is often recorded directly in a bound laboratory notebook. Furthermore, notebook records cannot be changed; only the interpretations or the conclusions from the recorded data are subject to revaluation. A laboratory notebook usually contains ruled graph paper and data is recorded in ink. A data entry is never erased or deleted. If a data error is made, the researcher strikes through the entry with a single line and enters the correct data.

A laboratory notebook can also serve as a legal document. For example, if a research chemist develops a product or process while working for a company, that company owns the rights. If another company "pirates" the product or process, the original company may be entitled to compensation. In court, a laboratory notebook may be presented as evidence to establish legal ownership.

In an introductory chemistry class, students are just beginning to develop lab skills. Accordingly, report forms are normally provided in introductory chemistry lab manuals. Moreover, the accurate recording of data in a report form is an important skill for students to develop. In many professions including healthcare, science, engineering, and technology, the completion of accurate report forms is a routine responsibility of the professional. Most healthcare workers are not involved in basic research, but have the important responsibility of recording clinical observations and patient vital sign data.

Conversely, scientific research requires a different set of skills as experiments are carefully planned in advance and the entire investigation, including data, is entered directly in a bound notebook. A suggestion for acquiring both skills—that is, writing report forms and keeping a laboratory notebook—is as follows:

(a) Initially, submit laboratory reports on the tear sheets provided showing all calculations and paying strict attention to significant digits and chemical formulas.

(b) After gaining skill in filling out report forms, the instructor may require students to submit a formal write-up to learn the task of keeping a laboratory notebook as shown in the following example.

(left notebook page – even numbered)

*TITLE OF EXPERIMENT**

Analysis of Vinegar

*REFERENCES**

Laboratory Manual for Introductory Chemistry: Concepts and Critical Thinking, 6/e, Pearson Education, Inc., Upper Saddle River (2013), pp. 229–240.

*OBJECTIVES**

List the principle objectives to be achieved by performing the experiment; for example,

1. To standardize a sodium hydroxide solution using solid potassium hydrogen phthalate.
2. To determine the molar concentration and mass/mass percent concentration of acetic acid in an unknown vinegar solution.
3. To gain proficiency in the laboratory techniques of titration.

*DISCUSSION**

Write a few short sentences which describe the theory of the experiment. For example, discuss acid-base neutralization, acid-base indicator changes, standardization of a solution, and the analysis of an unknown vinegar sample by titration.

Preparation of a Sodium Hydroxide Solution

We begin the experiment by diluting 6 *M* NaOH with water. Since the dilution of NaOH provides only an approximate concentration, it is necessary to determine the concentration precisely by standardization. To standardize NaOH, we will weigh crystals of potassium hydrogen phthalate, $KHC_8H_4O_4$ (abbreviated KHP). After dissolving the KHP crystals in water, we will titrate the acid solution with NaOH.

Titration of Acetic Acid in Vinegar

After standardizing the sodium hydroxide solution, we will determine the concentration of acetic acid in an unknown vinegar solution. A sample of vinegar will be titrated with NaOH to a phenolphthalein endpoint.

**Sections marked by an asterisk are to be completed before starting the experiment in laboratory.*

EQUIPMENT AND CHEMICALS*

- graduated cylinder
- 1000 mL Florence flask
- 125 mL Erlenmeyer flasks (3)
- buret stand
- 50 mL buret
- small funnel
- 10 mL pipet & bulb
- 100 mL beaker
- 150 mL beaker
- wash bottle with distilled water
- ~25 mL 6 *M* NaOH
- ~0.5 g samples KHP (3)
- ~50 mL unknown vinegar solution
- phenolphthalein indicator

PROCEDURE*

List each of the steps in the experiment. The purpose of writing out the procedure is to become familiar with the flow of activity in the experiment, and to be alerted to safety precautions.

A. *Preparation of a Sodium Hydroxide Solution*

1. Half-fill a 1000-mL Florence flask with ~500 mL of distilled water. Measure ~15 mL of 6 *M* NaOH into a graduated cylinder and pour the NaOH into the Florence flask. Stopper the flask, and carefully swirl to mix the solution.

2. Condition a buret with NaOH solution from the Florence flask. Use a small funnel and half-fill the buret with NaOH. Allow some solution to pass through the buret tip into a 150-mL beaker, and empty the remainder into the sink.

3. Close the stopcock, and fill the buret with NaOH solution from the Florence flask.

   Note: Carefully add NaOH solution to the funnel so as to not overfill the buret.

4. Label the 125-mL Erlenmeyer flasks #1, #2, and #3. Precisely weigh ~0.5 g of KHP into each of the flasks. Add ~25 mL of distilled water to each flask, and heat as necessary to dissolve the KHP crystals.

   Note: If a digital electronic balance is available, the Instructor may direct students to tare weigh the KHP samples.

5. Titrate three KHP samples as follows:
   - Drain NaOH through the tip of the buret to clear any air bubbles.
   - Position Erlenmeyer flask #1 under the buret as shown in Figure 20.1.
   - Record the initial buret reading (± 0.05 mL).
   - Add a drop of indicator to the flask.
   - Titrate with NaOH to a permanent endpoint while slowly swirling the flask.
   - Record the final buret reading (± 0.05 mL).

(left notebook page – even numbered)

*PROCEDURE (cont.)**

6. Refill the buret with NaOH solution, record the initial buret reading, add a drop of indicator to flask #2, titrate the KHP sample, and record the final buret reading.

7. Refill the buret with NaOH solution, record the initial buret reading, add a drop of indicator to flask #3, titrate the KHP sample, and record the final buret reading.

8. Calculate the molarity of the NaOH solution for each trial. Record the average molarity of NaOH in the Data Table of Procedure B.

   Note: The buret is conditioned before the first titration. It is not necessary to condition for subsequent titrations.

   Note: Save the NaOH in the Florence flask for Procedure B.

B. *Titration of Acetic Acid in Vinegar*

1. Obtain ~50 mL of vinegar solution in a dry 100-mL beaker. Record the unknown number in the Data Table.

2. Condition a pipet with unknown vinegar solution, and transfer a 10.0-mL sample into each 125-mL flask (see **Appendix E**). Add ~25 mL of distilled water into each flask.

   Note: It is not necessary to use dry flasks.

3. Titrate three vinegar samples as follows:
   - Position Erlenmeyer flask #1 under the buret.
   - Record the initial buret reading (± 0.05 mL).
   - Add a drop of indicator to the flask.
   - Titrate the vinegar sample to a permanent endpoint.
   - Record the final buret reading (± 0.05 mL).

4. Refill the buret with NaOH solution, record the initial buret reading, add a drop of indicator to flask #2, titrate the vinegar sample, and record the final buret reading.

5. Refill the buret with NaOH solution, record the initial buret reading, add a drop of indicator to flask #3, titrate the vinegar sample, and record the final buret reading.

6. Calculate the molarity of acetic acid, $HC_2H_3O_2$, in the unknown vinegar solution.

7. Convert the molarity of $HC_2H_3O_2$ (60.06 g/mol) to mass/mass percent concentration. Assume the density is 1.01 g/mL for the unknown vinegar solution.

   Note: When the titrations are complete, rinse the buret and all glassware with distilled water to remove traces of NaOH solution.

(right notebook page – odd numbered)

*CHEMICAL EQUATIONS**

Write a balanced chemical equation for each reaction in the experiment; for example,

A. Titration of KHP

$$KHP(s) + NaOH(aq) \rightarrow KNaP(aq) + H_2O(l)$$

B. Titration of $HC_2H_3O_2$

$$HC_2H_3O_2(aq) + NaOH(aq) \rightarrow NaC_2H_3O_2(aq) + H_2O(l)$$

*DATA TABLES**

Record all data directly in ink. If an error is made, draw a single line through the mistake and enter the correct value. To eliminate scientific bias, all data is considered valid unless there is an obvious mistake. If data is merely questionable, a statistical test must be applied before the data can be discarded.

A. Preparation of a Sodium Hydroxide Solution

mass of flask + KHP	_____ g	_____ g	_____ g
mass of flask	_____ g	_____ g	_____ g
mass of KHP	_____ g	_____ g	_____ g
final buret reading	_____ mL	_____ mL	_____ mL
initial buret reading	_____ mL	_____ mL	_____ mL
volume of NaOH	_____ mL	_____ mL	_____ mL

B. Titration of Acetic Acid in Vinegar

UNKNOWN # _____

volume of vinegar	10.0 mL	10.0 mL	10.0 mL
final buret reading	_____ mL	_____ mL	_____ mL
initial buret reading	_____ mL	_____ mL	_____ mL
volume of NaOH	_____ mL	_____ mL	_____ mL

(left notebook page – even numbered)

CALCULATIONS

Show all calculations associated with the values in the Data Table. Arithmetic operations, such as multiplication and division, are performed using a calculator and are not shown. Each calculation should be clearly presented using a consistent method of problem solving.

A. Preparation of a Standard Sodium Hydroxide Solution

#1 $$0.995\ \cancel{\text{g KHP}} \times \frac{1\ \cancel{\text{mol KHP}}}{204.23\ \cancel{\text{g KHP}}} \times \frac{1\ \text{mol NaOH}}{1\ \cancel{\text{mol KHP}}} = 0.00487\ \text{mol NaOH}$$

$$\frac{0.00487\ \text{mol NaOH}}{23.35\ \cancel{\text{mL}}\ \text{solution}} \times \frac{1000\ \cancel{\text{mL}}}{1\ \text{L}} = \frac{0.209\ \text{mol NaOH}}{1\ \text{L solution}} = 0.209\ M\ \text{NaOH}$$

#2 Show the calculation for trial **2**.

#3 Show the calculation for trial **3**.

B. Titration of Acetic Acid in Vinegar — UNKNOWN # ______

(a) molar concentration of $HC_2H_3O_2$

#1 $$39.70\ \text{mL NaOH} \times \frac{0.209\ \cancel{\text{mol NaOH}}}{1000\ \text{mL solution}} \times \frac{1\ \text{mol HC}_2\text{H}_3\text{O}_2}{1\ \cancel{\text{mol NaOH}}} = 0.00830\ \text{mol HC}_2\text{H}_3\text{O}_2$$

$$\frac{0.00830\ \text{mol HC}_2\text{H}_3\text{O}_2}{10.0\ \cancel{\text{mL}}\ \text{solution}} \times \frac{1000\ \cancel{\text{mL}}}{1\ \text{L}} = \frac{0.830\ \text{mol HC}_2\text{H}_3\text{O}_2}{1\ \text{L solution}} = 0.830\ M\ \text{HC}_2\text{H}_3\text{O}_2$$

#2 Show the calculation for trial **2**.

#3 Show the calculation for trial **3**.

(b) mass/mass percent concentration of $HC_2H_3O_2$

#1 $$\frac{0.833\ \cancel{\text{mol HC}_2\text{H}_3\text{O}_2}}{1000\ \cancel{\text{mL solution}}} \times \frac{60.06\ \text{g HC}_2\text{H}_3\text{O}_2}{1\ \cancel{\text{mol HC}_2\text{H}_3\text{O}_2}} \times \frac{1\ \cancel{\text{mL solution}}}{1.01\ \text{g solution}} \times 100\% = 4.95\%$$

#2 Show the calculation for trial **2**.

#3 Show the calculation for trial **3**.

RESULTS

Summarize the experimental results as shown in the following example. Report the average value for the unknown along with the unknown number. Upon completion of the experiment, sign and date the results to authenticate your work.

Note: In more advanced courses, you will analyze your results, suggest sources of error, and apply statistical tests to disregard data that is questionable.

molarity of NaOH	0.209 *M*	0.211 *M*	0.206 *M*
average molarity of NaOH		0.209 *M*	
molarity of $HC_2H_3O_2$	0.830 *M*	0.835 *M*	0.834 *M*
average molarity of $HC_2H_3O_2$		0.833 *M*	
mass/mass percent $HC_2H_3O_2$	4.95%	5.02%	5.03%
average m/m % $HC_2H_3O_2$		5.00%	

UNKNOWN # ______

Signature *I. M. Student* **Date** *10/15/11*

CONCLUSIONS

In advanced courses, you will state conclusions based on the results. For this introduction to the laboratory notebook, simply answer the questions in the postlab assignment. The postlab assignment reinforces the theory and practice in the experiment.

APPENDIX

I

Glossary

–A–

abscissa The horizontal axis (x-axis) on a graph.
activity series A relative order of metals arranged in a list according to their ability to undergo reaction. A metal higher in the series will displace another metal from its aqueous solution.
actual yield The amount of product experimentally obtained from a reaction.
alkali metal Any Group IA/1 element in the periodic table, excluding hydrogen.
alkaline earth metal Any Group IIA/2 element in the periodic table.
anhydrous compound A compound that does not contain water.
anion Any negatively charged ion.
aqueous solution A solution of a substance dissolved in water.
atmospheric pressure The pressure exerted by the air molecules in Earth's atmosphere; atmospheric pressure is measured with a barometer.
"atomic fingerprint" The unique line spectrum that is characteristic of a given element and can be used for identification.
Avogadro's number (*N*) The value that corresponds to the number of carbon atoms in 12.01 g of carbon; 6.02×10^{23} particles.

–B–

Balmer formula A mathematical formula for calculating the wavelength of light emitted from an excited hydrogen atom when the electron drops to the second energy level.
biochemistry The study of compounds derived from plants and animals.

–C–

carbonyl group The C=O group, which is present in aldehydes, ketones, carboxylic acids, esters, and amides.
catalyst A substance that speeds up a chemical reaction.
cation Any positively charged ion.
centimeter (cm) A metric unit of length; 100 cm = 1 m.
centrifuge An instrument that spins test tubes in order to separate a precipitate from solution. The act of rapidly spinning a test tube in order to separate a precipitate from solution.
change of state The conversion from one physical state to another; for example, the change in a substance from a liquid to a solid.

chemical change A modification of a substance that alters its chemical composition.
chemical property A characteristic of a substance that cannot be observed without changing its chemical formula.
chemistry The branch of science that studies the composition and properties of matter.
chromatography A method for separating a mixture into its components as a result of a varying attraction of compounds for a mobile solvent on a stationary solid.
class of compounds A family of compounds in which all the members have the same structural feature (that is, an atom or group of atoms) and similar chemical properties.
combined gas law The pressure exerted by a gas is inversely proportional to its volume and directly proportional to its Kelvin temperature.
compound A pure substance that can be broken down into two or more simpler substances by chemical reaction.
condition The act of rinsing glassware (e.g., a graduated cylinder, pipet, or buret) with a sample liquid to avoid dilution by water on the inside surface.
continuous spectrum A broad uninterrupted band of radiant energy.
coprecipitation A term that refers to a precipitate containing impurities that are usually soluble.
covalent bond A bond characterized by the sharing of one or more pairs of valence electrons.

–D–

Dalton's law of partial pressures The pressure exerted by a mixture of gases is equal to the sum of the pressures exerted by each gas in the mixture.
decant The process of pouring a liquid from one container into another; for example, pouring the supernate from one test tube into a second test tube.
density (*d*) The amount of mass in a unit volume of matter; for example, g/mL.
digestion The process of heating a precipitate in aqueous solution to develop larger particles that are easier to filter and free of impurities.
dissociation The process of an ionic compound dissolving in water and separating into positive and negative ions.
double bond A bond composed of two electron pairs shared between two atoms. A double bond is represented by two dashes between the symbols of the two atoms.

–E–

electron dot formula A diagram of a molecule in which each atom is represented by its chemical symbol surrounded by two dots for each pair of bonding or nonbonding electrons.
element A pure substance that cannot be broken down any further by ordinary chemical reaction.
empirical formula The chemical formula of a compound that expresses the simplest whole number ratio of atoms of each element in a molecule, or ions in a formula unit.
endpoint The stage in a titration when the indicator changes color.
experiment A scientific procedure for collecting data and recording observations under controlled conditions.
experimental conditions The conditions of temperature and pressure at which a gas sample is collected; not usually STP.

–F–

filtrate The solution that passes through filter paper in a filtration operation.
firing to red heat Heating a crucible or other porcelain object until it glows red.
flame test A means of identifying a substance by observing the characteristic color it emits when placed in a hot flame.
freezing point The temperature at which a liquid substance crystallizes and forms a solid.
frequency The number of times a light wave travels a complete cycle in one second.
functional group An atom or group of atoms that characterizes a class of compounds, and contributes to their similar physical and chemical properties.

–G–

gram (g) A metric unit of mass; 1000 g = 1 kg.
group A vertical column in the periodic table; a family of elements having similar properties.

–H–

halide A negatively charged Group VIIA/17 element; for example, bromide, chloride, or iodide.
halogen A Group VIIA/17 element; for example, chlorine, bromine, or iodine.
heating to constant weight A repeated process of heating, cooling, and weighing until the mass readings for an object are constant, or agree closely.
heterogeneous mixture Matter having an indefinite composition and properties that can vary within the sample.
homogeneous mixture Matter having a definite composition but properties that can vary from sample to sample; examples include alloys, solutions, and gas mixtures.
hydrate A substance that contains a specific number of water molecules attached to a formula unit in a crystalline compound.
hydrocarbon A compound containing only hydrogen and carbon.
hydrocarbon derivative A compound containing carbon, hydrogen, and another element such as oxygen, nitrogen, or a halogen.
hypothesis A tentative proposal of a scientific principle that attempts to explain the meaning of a set of data collected in an experiment.

–I–

immiscible A term that refers to liquids that do not dissolve in one another and separate into two layers.
indicator A chemical substance that undergoes a color change according to the pH of a solution; for example, phenolphthalein is colorless below pH 9 and pink above pH 9.
ionization The process of a polar molecular compound dissolving in water and forming positive and negative ions.
isomers Compounds with the same molecular formula but with different structural formulas. Isomers have different physical and chemical properties.

–L–

light A specific term that refers to the portion of the radiant energy spectrum that is visible; that is, violet through red. A general term that refers to all forms of radiant energy.

***like dissolves like* rule** The general principle that solubility is greatest when the polarity of the solute and solvent are similar.

line spectrum The narrow bands of light observed through a spectroscope that are emitted from excited atoms in a gas discharge tube.

–M–

mass The amount of matter in a sample. Mass is independent of Earth's gravitational attraction and is the quantity measured with a laboratory balance.

mass/mass percent concentration (m/m %) A solution concentration expression that relates the mass of solute in grams dissolved in each 100 grams of solution.

$$\frac{\text{mass of solute}}{\text{mass of solution}} \times 100\% = \text{m/m }\%$$

melting point The temperature at which a solid substance melts and forms a liquid.

meniscus A clear lens at the surface of a liquid inside a piece of narrow glassware, such as a graduated cylinder, pipet, or buret.

metric system A decimal system of measurement using prefixes and a basic unit to express physical quantities such as length, mass, and volume.

milliliter (mL) A metric unit of volume; 1000 mL = 1 L.

miscible A term that refers to liquids that dissolve completely in one another.

mobile phase A term that refers to the solvent that travels up a paper chromatogram by capillary action.

molar concentration (*M*) A solution concentration expression that relates the moles of solute dissolved in each liter of solution; also referred to as molarity.

$$\frac{\text{moles of solute}}{\text{liters of solution}} = M$$

molar mass (MM) The mass of 1 mole of any substance expressed in grams. (The individual particles that compose the substance may be atoms, molecules, or formula units.)

molar volume The volume occupied by 1 mole of any gas at STP; at 0°C and 760 mm Hg, the volume of 1 mole of any gas is 22.4 L (22,400 mL).

mole (mol) The amount of substance that contains Avogadro's number of particles; that is, an amount of substance that contains 6.02×10^{23} particles.

molecular formula The chemical formula of a compound that expresses the actual number of atoms of each element in a molecule.

monolayer A thin film layer of organic molecules on the surface of water; the monolayer is only 1 molecule thick.

–N–

nanometer (nm) A unit of length used to express wavelengths of light; a unit of length equal to one-billionth of a meter.

net ionic equation A chemical equation that portrays an ionic reaction after spectator ions have been canceled from the total ionic equation. The net ionic equation shows only those species undergoing a change during a chemical reaction.
nonpolar "tail" The portion of a long organic molecule having nonpolar bonds.

–O–

octet rule The statement that an atom tends to bond in such a way so as to acquire eight electrons in its outer shell. A hydrogen atom is an exception to the rule and acquires only two valence electrons.
ordinate The vertical axis (y-axis) on a graph.
organic chemistry The study of compounds that contain one or more carbon atoms.
origin The point of intersection of the horizontal and vertical axes on a graph.
oxidation A chemical process characterized by the loss of electrons. A process in which a substance undergoes an increase in oxidation number.
oxidation number A positive or negative value assigned to an atom in a substance according to a set of rules. A value that indicates whether an atom is electron poor or electron rich compared to the free atom. Metals and nonmetals in the free state have an oxidation number of zero.
oxidizing agent A substance that causes the oxidation of another substance in a redox reaction. The substance that is reduced in a redox reaction.

–P–

percent yield The actual yield compared to the theoretical yield expressed as a percent.
period A horizontal row in the periodic table; a series of elements with properties that vary from metallic to nonmetallic.
periodic law The properties of the elements recur in a repeating pattern when arranged according to increasing atomic number.
photon A particle of light that corresponds to a unit of radiant energy. A photon may also be referred to as a quantum (*pl.,* quanta).
physical change A modification of a substance that does not alter its chemical composition; for example, a change in physical state.
physical property A characteristic of a substance that can be observed without changing its chemical formula.
physical state The condition of a substance existing as a solid, liquid, or gas.
polar "head" The portion of a long organic molecule having polar bonds.
precipitate (ppt) An insoluble solid substance produced from a reaction in aqueous solution.
product A substance resulting from a chemical reaction.

–Q–

qualitative analysis A systematic procedure for the separation and identification of cations, anions, or other substances present in a sample.

–R–

reactant A substance undergoing a chemical reaction.
redox reaction A chemical reaction that involves electron transfer and causes reduction of one species and oxidation of another.
reducing agent A substance that causes the reduction of another substance in a redox reaction. The substance that is oxidized in a redox reaction.
reduction A chemical process characterized by the gain of electrons. A process in which a substance undergoes a decrease in oxidation number.
R_f value The ratio of the distance traveled by a sample component compared to the distance traveled by the solvent.
Rydberg equation A general mathematical equation for calculating the wavelength of light emitted from an excited hydrogen atom when the electron drops to any lower energy level.

–S–

saturated hydrocarbon A hydrocarbon containing only single bonds.
science The methodical exploration of nature and the logical explanation of the observations.
scientific method A systematic investigation that involves performing an experiment, proposing a hypothesis, testing the hypothesis, and stating a theory or law that explains a scientific principle.
single bond A bond composed of one electron pair shared between two atoms. A single bond is represented by a dash between the symbols of the two atoms.
solute The component of a solution that is the lesser quantity.
solvent The component of a solution that is the greater quantity.
solvent front The leading edge of the solvent, which travels from the bottom of the developing chamber to the upper portion of the chromatogram.
spectator ions Those ions in aqueous solution that do not participate in a reaction, and do not appear in the net ionic equation.
standard conditions See standard temperature and pressure.
standard solution A solution whose concentration has been established precisely (usually by titration to 3 or 4 significant digits).
standard temperature and pressure (STP) A temperature of 0°C and a pressure of 1 atm. A temperature of 273 K and a pressure of 760 mm Hg for a gas.
stationary phase A term that refers to the moisture that is strongly adsorbed onto a paper chromatogram and is not free to travel.
stoichiometry The relationship of quantities (i.e., mass of substance or volume of gas) in a chemical reaction according to the balanced chemical equation.
strong electrolyte An aqueous solution that is a good conductor of electricity and produces a bright glow from a light bulb in a conductivity apparatus.
structural formula A diagram of a molecule or polyatomic ion that shows the chemical symbol of each atom and a dash representing each pair of bonding electrons.
sublimation The direct change of state from a solid to a gas without forming a liquid. Conversely, the direct change of state from a gas to a solid is called deposition.
substance Matter having constant composition with definite and predictable properties.
supernate The solution above a precipitate after insoluble particles are separated from solution.

supersaturated solution A solution containing more solute than can ordinarily dissolve at a given temperature. A supersaturated solution is unstable and the excess solute will crystallize from solution if a seed crystal is added.
surface area Specifically, the region occupied by a single layer of organic molecules floating on water; the formula for calculating the surface area of a circle is $\pi d^2/4$.

–T–

tare weighing A procedure for obtaining the mass of a sample *directly* by placing a container on an electronic balance and setting the balance to zero. Second, add a sample to the container and record the mass of sample directly.
theoretical yield The amount of product that is calculated from a given amount of reactant.
theory An extensively tested proposal of a scientific principle that explains the behavior of nature. A theory offers a model, for example the atomic theory, to describe nature.
titration A laboratory procedure for delivering a measured volume of solution using a buret.
total ionic equation A chemical equation that portrays highly ionized substances in the ionic form and slightly ionized substances in the nonionized form.
triple bond A bond composed of three electron pairs shared between two atoms. A triple bond is represented by three dashes between the symbols of the two atoms.

–U–

uncertainty A term that refers to the degree of inexactness in an instrumental measurement; for example, ± 0.05 cm, ± 0.001 g, ± 0.5 mL, ± 0.5°C, or ± 1 s.
unsaturated hydrocarbon A hydrocarbon containing a carbon-carbon double or triple bond.

–V–

valence electrons The electrons in the outermost *s* and *p* sublevels of an atom that form chemical bonds.
vapor pressure The pressure exerted by gaseous vapor above a liquid in a closed container when the rates of evaporation and condensation are equal; for example, the pressure exerted by water vapor above liquid water.
visible spectrum Light energy that is observed as violet, blue, green, yellow, orange, and red; the region in the radiant energy spectrum from approximately 400–700 nm.
volume by displacement A technique for determining the volume of a sample by measuring the volume of water it displaces.

–W–

water of crystallization The number of water molecules bound to a formula unit in a hydrate; also called *water of hydration*.
wavelength (λ) The distance a light wave travels to complete one cycle.
weak electrolyte An aqueous solution that is a poor conductor of electricity and produces a dim glow from a light bulb in a conductivity apparatus.
weighing by difference A procedure for obtaining the mass of a sample *indirectly* by first weighing a container and then weighing the container with the sample.

APPENDIX

J

Answers to Prelaboratory Assignments

Experiment 1 — *Introduction to Chemistry*

1. See the Glossary, Appendix I.
2. Refer to the diagrams of Common Laboratory Equipment, pages 4–5.
3. Do the Prelaboratory Assignment *before* coming to lab. Instructions (b)–(j) are true.
4. Flush immediately with tap water and notify the Instructor.
5. • Wear safety goggles; be careful when handling chemicals and solutions.
   • Handle glassware carefully, as it is easily broken and can cause cuts.
   • Dispose of chemical waste in the designated containers.

Experiment 2 — *Instrumental Measurements*

1. See the Glossary, Appendix I.
2. 8.8 cm, 14.0 cm
3. 18.75 = 19 cm^3
4. 7.15 cm; 10.00 cm
5. 18.55 = 18.6 cm^3
6. 14.0 mL, 83.5 mL
7. 30.0 °C, 2.5 °C
8. • Wear safety goggles; be careful when using the laboratory burner.
   • Handle the boiling waterbath carefully, as it can cause burns.
   • Handle the thermometer carefully, as it is easily broken and can cause cuts.

Experiment 3 — *Density of Liquids and Solids*

1. See the Glossary, Appendix I.
2. 0.799 g/mL
3. 54.0 mL, 62.5 mL
4. 1.2 g/mL
5. 5.40 cm, 4.25 cm
6. 8.90 g/cm^3
7. 0.00170 cm (1.70×10^{-3} cm)
8. • Wear safety goggles; be careful handling the 10-mL pipet.
   • Dispose of solvents in the organic waste container.

Experiment 4 — *Freezing Points and Melting Points*

1. See the Glossary, Appendix I.
2. Hot tap water leaves mineral deposits on glassware.
3. ~40 °C
4. ~75 °C
5. 65.0 °C
6. The freezing point is the flat plateau on the cooling curve. The freezing point is determined by extending a straight line from the plateau back to the vertical axis.

7. Place the test tube with solid paradichlorobenzene in a hot waterbath. After the compound melts, remove the thermometer and wipe off the residue with a paper towel.
8. A melting point is time-consuming when heating a compound slowly. Thus, first determine an approximate melting point by heating rapidly. Second, heat rapidly to within a few degrees of the approximate melting point, and then 1 °C per minute to get an accurate melting point.
9. 75–85 °C
10. 79.0–80.5 °C
11. If the compound appears to liquefy, the problem may be:
    (1) The temperature of the waterbath is higher than the melting point of the compound.
    (2) The capillary may not be sealed completely, and water may be leaking into the tube.
12. • Wear safety goggles; be careful when using the laboratory burner.
    • Do not heat the test tube directly because paradichlorobenzene is flammable.
    • Do not pour out the liquid paradichlorobenzene, as it is used for repeated trials.
    • Handle the thermometer carefully, and do not attempt to remove a thermometer frozen in solid paradichlorobenzene. (Report a broken thermometer immediately to the Instructor.)
    • Dispose of chemical waste in the designated container.

Experiment 5 — *Physical Changes and Chemical Changes*

1. See the Glossary, Appendix I.
2. (a) phys; (b) phys; (c) phys; (d) phys; (e) phys; (f) chem; (g) phys; (h) phys
3. (a) chem; (b) phys; (c) phys; (d) chem; (e) chem; (f) phys; (g) chem; (h) chem
4. The boiling chip prevents "bumping" that can eject flammable liquid from the test tube.
5. All of the following indicate a chemical change:
   (1) a solution releases gas bubbles; (2) a solution forms an insoluble solid; (3) a solution undergoes a permanent color change; (4) a solution releases or absorbs energy.
6. A gas is indicated if there is fizzing, bubbling, or an odor is observed.
7. • Wear safety goggles; keep flammable organic liquids away from a burner flame.
   • Handle the thermometer carefully, as it is easily broken and can cause cuts. (Report a broken thermometer immediately to the Instructor.)
   • Dispose of chemical waste in the designated container.

Experiment 6 — *"Atomic Fingerprints"*

1. See the Glossary, Appendix I.
2. (a) increasing wavelength: V > B > G > Y > O > R (Red has the longest wavelength.)
   (b) increasing frequency: R > O > Y > G > B > V (Violet has the highest frequency.)
   (c) increasing energy: R > O > Y > G > B > V (Violet has the highest energy.)
3. 430 nm, 480 nm, 650 nm
4. violet, blue-green, red
5. We can calculate the wavelength of the emission line when electrons in a hydrogen atom drop from energy level 6 to 2 using the Balmer formula:

$$\frac{1}{\lambda} = \frac{1}{91\text{ nm}}\left(\frac{1}{2^2} - \frac{1}{6^2}\right) = \frac{0.22}{91\text{ nm}}$$

$$\lambda = \frac{91\text{ nm}}{0.22} = 413.6\text{ nm} = 410\text{ nm}$$

   Since the 6 to 2 drop is not common, the spectral line is weak. However, if you closely observe the emission spectrum of hydrogen, you may see a faint violet line at 410 nm.
6. When 1 electron drops from energy level 6 to 2, 1 photon is emitted.
   When 10 electrons drop from energy level 6 to 2, 10 photons are emitted.
7. • Do not drop the hand spectroscope as the wavelength scale can misalign.
   • To avoid a burn, do not touch a hot gas discharge tube in the power supply.
   • To avoid a shock, do not touch the discharge tube while the power supply is turned on.

Experiment 7 — *Families of Elements*

1. See the Glossary, Appendix I.
2. alkali: Li, K, Na; alkaline earth: Ba, Ca, Sr
3. (a) Ba; (b) Ca; (c) Li; (d) K; (e) Na; (f) Sr
4. When performing a flame test, the end of the wire is placed at the tip of the burner flame.
5. Sodium impurity gives a weak, yellow, flame test. An aqueous sodium solution gives a strong, long-lasting, flame test.
6. bromide, chloride, iodide
7. Water and hexane are *immiscible* and separate into two layers.
8. The halide test is observed in the *upper layer*.
9. (a) bromide; (b) chloride; (c) iodide
10. • Wear safety goggles; be careful when using the burner and performing a flame test.
    • Handle acids carefully, and avoid breathing the vapors of conc hydrochloric acid.
    • Dispose of chemical waste in the designated container.

Experiment 8 — *Identifying Cations in Solution*

1. See the Glossary, Appendix I.
2. Ba^{2+}, Ca^{2+}, and Mg^{2+}
3. When performing a flame test, the end of the wire is placed at the tip of the burner flame.
4. A blue spot indicates a basic solution. If the spot is red, add a few more drops of NaOH.
5. (a) Ba^{2+}; (b) Ca^{2+}; and (c) Mg^{2+}
6. Tap water contains ions that can interfere with the analysis.
7. All three cations (Ba^{2+}, Ca^{2+}, and Mg^{2+}) are present.
8. Mg^{2+} is present; Ba^{2+} and Ca^{2+} are absent.
9. • Wear safety goggles; be careful when using the burner and performing a flame test.
   • Avoid contact with HCl and NaOH. If contacted, wash immediately with water.
   • Balance the centrifuge before operating.
   • Dispose of chemical waste in the designated container.

Experiment 9 — *Identifying Anions in Solution*

1. See the Glossary, Appendix I.
2. I^-, Cl^-, SO_4^{2-}
3. A red spot indicates an acidic solution. If the spot is blue, add a few more drops of HNO_3.
4. (a) I^- ; (b) Cl^-; and (c) SO_4^{2-}
5. Tap water contains ions that can interfere with the analysis.
6. All three anions (I^-, Cl^-, SO_4^{2-}) are present.
7. SO_4^{2-} is present; I^- and Cl^- are absent.
8. • Wear safety goggles; avoid contact with HNO_3, NH_4OH, and $AgNO_3$. If contacted, wash the area immediately with water.
   • Dispose of chemical waste in the designated container.

Experiment 10 — *Analysis of a Penny*

1. See the Glossary, Appendix I.
2. (a) combination, (b) decomposition, (c) single replacement, (d) double replacement, (e) neutralization
3. (a) a gas is released; (b) a precipitate is produced; (c) a permanent color change is observed; (d) an energy change is noted, such as heat or light.
4. (a) (g); (b) (l); (c) (s); (d) (aq); (e) (s); (f) Δ; (g) *NR*
5. (a) colorless; (b) pink
6. The 1980 mint date cannot be used as it is a "copper penny."
7. (a) 2.5% Cu; (b) 97.48% Zn

8. • Wear safety goggles; be careful when using the laboratory burner.
   • Handle chemicals carefully, and avoid breathing the vapors of hydrochloric acid.
   • Dispose of chemical waste in the designated container.

Experiment 11 — *Determination of Avogadro's Number*

1. See the Glossary, Appendix I.
2. –COOH.
3 The "lens" disappears.
4. When the "lens" remains for ~30 seconds.
5. • Error is introduced by delivering drops of solution that are too large or vary in size.
   • It is important that the surface of the watchglass is clean. Touching the watchglass surface with your fingers can leave a trace of skin oil that leads to erroneous results.
6. (a) 113 cm^2; (b) 5.4×10^{16} molecules; (c) 8.5×10^{-8} mol; (d) 6.4×10^{23} molecules/mol
7. • Wear safety goggles; be careful with the dropper pipet as it is sharp and delicate.
   • Avoid breathing the organic hexane vapor.
   • Keep the flammable hexane liquid and vapor away from a laboratory burner flame.
   • Dispose of chemical waste in the designated container.

Experiment 12 — *Empirical Formulas of Compounds*

1. See the Glossary, Appendix I.
2. The empty crucible and cover are fired to red heat to burn off impurities and establish a constant weight.
3. The suggested periods for heating and cooling are general guidelines.
4. Igniting magnesium in air produces magnesium oxide and magnesium nitride. Adding distilled water to the crucible decomposes magnesium nitride and releases ammonia gas. (The crucible contents fizz, and the odor of ammonia gas is noticeable.) Reheating the crucible and contents converts magnesium hydroxide to magnesium oxide.
5. If the magnesium has reacted completely, there are no sparks when lifting the crucible cover.
6. The copper wire has reacted completely when there is no longer traces of sulfur in the crucible. If there is any doubt, heat the crucible to constant weight.
7. $Mg_{0.00991}O_{0.00988} = MgO$
8. • A hot crucible on the balance causes a weighing error, and mass readings will be low.
   • If white smoke escapes from the crucible, the mass readings will be low.
   • Sulfur has a tendency to "creep" out of the crucible during the firing. The excess sulfur must be driven off by heating, or the mass readings will be high.
9. • Wear safety goggles; be careful when using the laboratory burner.
   • Carefully lift the crucible cover with tongs to check the progress of the reaction.
   • Before firing to red heat, set the crucible on the lab bench and strike sharply with a pencil. A crucible with a hairline crack gives a dull ring.
   • A crucible that glows red has a temperature near 1100 °C. Below this temperature, a crucible may not glow red, but can cause a burn.
   • The ignition of magnesium ribbon is strongly exothermic and can crack a crucible. A few porcelain chips in the bottom of the crucible minimizes this problem, and can be weighed with the crucible and cover.
   • Heating copper and sulfur produces pungent sulfur dioxide gas. Avoid breathing the gas, and perform the reaction under a fume hood.
   • Dispose of chemical waste in the designated container.

Experiment 13 — *Analysis of Alum*

1. See the Glossary, Appendix I.
2. The popped corn *weighs less* than the unpopped kernels because steam escapes. Similarly, the mass of the anhydrous compound *weighs less* than the hydrate because steam escapes.

3. The hydrate is completely decomposed when moisture inside the beaker is gone. The sample will change in appearance from crystals (*before heating*) to powder (*after heating*).
4. A warm beaker radiates heat and warms the air around the balance pan. This warm air rises and lifts the balance pan causing a light mass reading.
5. • Not allowing the beaker to cool causes a light weighing, which gives *high results*.
   • Moisture on the watchglass causes a heavy weighing, which gives *low results*.
   • Incomplete heating of the hydrate causes a heavy weighing, which gives *low results*.
   • Overheating the hydrate can decompose the anhydrous compound, which gives *high results*.
6. (0.420 g / 0.932 g) x 100% = 45.2%
   The experimental result of 45.2% agrees with the theoretical value of 45.58%.
7. (30.5 g / 18.02 g/mol = 1.69 mol H_2O); (69.5 g / 164 g/mol = 0.424 mol AC)
   After dividing mol H_2O by mol AC (1.69 mol / 0.424 mol), the water of crystallization is found to be 4, and the formula is $AC \cdot 4H_2O$.
8. • Wear safety goggles; be careful when using the laboratory burner.
   • Heat the watchglass gently above a *low* burner flame to avoid breakage.
   • When weighing the beaker and watchglass, handle carefully to avoid breakage.
   • Dispose of chemical waste in the designated container.

Experiment 14 — *Decomposing Baking Soda*

1. See the Glossary, Appendix I.
2. The baking soda is decomposed completely when carbon dioxide gas is no longer produced, and the water level in the beaker remains constant.
3 Yes, some errors can lead to high results, giving a percent yield greater than 100%.
4. • Heating the baking soda mixture insufficiently gives high results.
   • Weighing a test tube with traces of moisture gives a heavy mass reading.
   • Weighing a test tube while warm gives a light mass reading.
5. 0.991 g; (0.991 g / 0.981 g) x 100% = 101%
6. (0.880 g / 1.473 g) x 100% = 60.0%
7. • Wear safety goggles; be careful when using the laboratory burner.
   • Heat the samples in a test tube slowly and carefully.
   • Avoid pinching the rubber tubing from the test tube to the Florence flask.
   • Dispose of chemical waste in the designated container.

Experiment 15 — *Precipitating Calcium Phosphate*

1. See the Glossary, Appendix I.
2. Digestion coagulates the precipitate, which speeds the rate of filtration.
3. If precipitate appears in the filtrate, you may have to recycle the filtrate through a second disk of weighed filter paper (ask the Instructor). If two filtrations are required, find the mass of precipitate from each filtration and add the two masses together.
4. The purpose of the rubber policeman is to clean the sides of the beaker in order to transfer all precipitate to the filter paper.
5 Yes, some errors can lead to high results, giving a percent yield greater than 100%.
6. • Incomplete precipitation gives low results; coprecipitation gives high results.
   • Particles of precipitate in the filtrate give low results.
   • Weighing filter paper with precipitate before it is completely dry gives high results.
   • Weighing warm filter paper before it cools gives low results.
7. 0.480 g; (0.484 g / 0.480 g) x 100% = 101%
8. (0.456 g / 0.700 g) x 100% = 65.1%
9. • Wear safety goggles; be careful when using the laboratory burner.
   • Handle all glassware carefully.
   • Dispose of chemical waste in the designated container.

Experiment 16 — *Generating Hydrogen Gas*

1. See the Glossary, Appendix I.
2. The graduated cylinder remains full of water after inverting due to atmospheric pressure.
3. If the magnesium weighs more than 0.09 g, the metal produces more than 100 mL of gas, which exceeds the capacity of the graduated cylinder.
4. The reaction is complete when the metal disappears and gas bubbles are no longer observed.
5. A "wet gas" is collected over water and contains water vapor; a "dry gas" gas does not. To obtain the partial pressure of a "dry gas," subtract the vapor pressure of water from the atmospheric pressure at experimental conditions.
6. The partial pressure of a "dry gas" equals the atmospheric pressure minus vapor pressure. Table 16.1 lists the vapor pressure at 25 °C as 24 mm Hg. Therefore, the partial pressure of "dry" hydrogen gas is 766 mm Hg – 24 mm Hg = 742 mm Hg.
7. 79.2 mL H_2 at STP; 0.00327 mol H_2 at STP; 24.2 L/mol
8. The major sources of error in this experiment include:
   - a large air bubble in the graduated cylinder, after inverting the cylinder filled with water
   - incomplete reaction of the metal
   - misreading the meniscus in the graduated cylinder
9.
   - Wear safety goggles; be careful with the glassware when collecting the gas.
   - Handle hydrochloric acid carefully, and avoid breathing the vapor.
   - Handle the thermometer carefully, as it is easily broken. (Report a broken thermometer immediately to the Instructor; mercury vapor is hazardous.)
   - Dispose of chemical waste in the designated container.

Experiment 17 — *Generating Oxygen Gas*

1. See the Glossary, Appendix I.
2. Manganese(IV) oxide is added as a catalyst for a safe and rapid decomposition of $KClO_3$.
3. A $KClO_3$ sample is decomposed completely when oxygen gas is no longer produced, and the water level in the beaker remains constant. (After the burner is removed, the water level in the beaker will decrease slightly as the carbon dioxide gas cools.)
4. Since the density of water is 1.00 g/mL, the volume of water corresponds to the mass of water; that is, 345.0 g H_2O = 345.0 mL H_2O. The volume of oxygen gas produced equals the volume of water displaced; that is, 345.0 mL H_2O = 345.0 mL O_2.
5. A "wet gas" is collected over water and contains water vapor; a "dry gas" gas does not. To obtain the partial pressure of a "dry gas," subtract the vapor pressure of water from the atmospheric pressure at experimental conditions.
6. The partial pressure of a "dry gas" equals the atmospheric pressure minus vapor pressure. Table 17.1 lists the vapor pressure at 23 °C as 21 mm Hg. Therefore, the partial pressure of "dry" oxygen gas is 765 mm Hg – 21 mm Hg = 744 mm Hg.
7. 248 mL O_2 at STP; 0.905 g $KClO_3$; 90.5%
8.
   - Heating a potassium chlorate mixture insufficiently gives high results.
   - Weighing a warm test tube gives a light mass reading.
9.
   - Wear safety goggles; be careful when using the laboratory burner.
   - Heat the samples in a test tube slowly and carefully.
   - A $KClO_3$ sample must not contact the rubber stopper in the test tube.
   - Avoid pinching the rubber tubing from the test tube to the Florence flask.
   - Dispose of chemical waste in the designated container.

Experiment 18 — *Molecular Models and Chemical Bonds*

1. See the Glossary, Appendix I.
2. The number of valence electrons corresponds to the group number of the element in the periodic table; thus,

(a)	C = 4 (Group IVA/14)	(b)	H = 1 (Group IA/1)
(c)	O = 6 (Group VIA/16)	(d)	N = 5 (Group VA/15)
(e)	Cl = 7 (Group VIIA/17)	(f)	Br = 7 (Group VIIA/17)

3.

(a)	single bond (one e– pair)	(b)	double bond (two e– pairs)
(c)	triple bond (three e– pairs)	(d)	hydrogen atom
(e)	carbon atom	(f)	oxygen atom
(g)	chlorine atom	(h)	bromine atom
(i)	iodine atom	(j)	nitrogen atom

4. (a) I—Br

   (b)

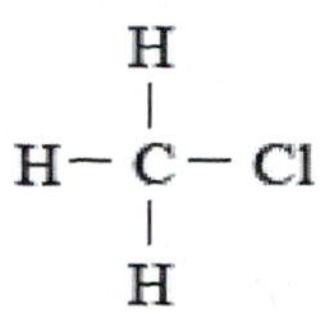

   (c)

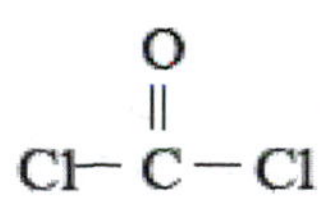

5. (a)

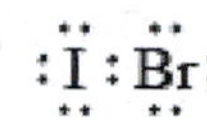

   (b)

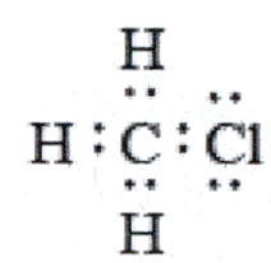

   (c)

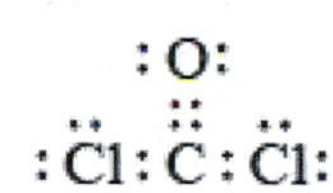

6. (a) 14 e–; (b) 14 e–; (c) 24 e–

Experiment 19 — *Analysis of Saltwater*

1. See the Glossary, Appendix I.
2. The dark brown iodine crystal will give a faint yellow color to the solvent.
3. Two miscible solvents, upon shaking, form one homogeneous solution.
   Two immiscible solvents, after shaking, separate into two layers.
4. Heating tap water leaves carbonate deposits on glassware; distilled water does not.
5. (a) 5.00%
   (b) 0.874 *M*
6. • Pipetting is the main source of error. A 10.0-mL sample of saltwater should weigh slightly more than 10.0 g, for example 10.214 g. Check your data, and pipet again if necessary.
   • An evaporating dish that has moisture on the bottom gives rise to high results.
   • Not allowing the evaporating dish to cool completely before weighing gives high results.
7. • Wear safety goggles; be careful when using the laboratory burner, thermometer, and pipet.
   • Keep organic liquids and vapors away from a burner flame.
   • Avoid breathing the vapor from the organic solvents.
   • Dispose of chemical waste in the designated container.

Experiment 20 — *Analysis of Vinegar*

1. See the Glossary, Appendix I.
2. (a) 0.50 mL (b) 31.35 mL
3. You are nearing the endpoint in a titration when flashes of indicator color persist longer.
4. Only 1 drop of NaOH is required to change the indicator permanently at the endpoint.
5. The volume of NaOH required for an endpoint varies for each trial, depending on the mass of the KHP sample.
6. The volume of NaOH required for an endpoint should be about the same for each trial because the amount of acetic acid is the same in each vinegar sample.
7. 0.206 *M* NaOH
8. (a) 0.594 *M* $HC_2H_3O_2$; (b) 3.53% $HC_2H_3O_2$
9. (c), (d), (e), and (g) are serious errors.
10. • Wear safety goggles; be careful when using the laboratory burner, pipet, and buret.
    - Add NaOH carefully to the buret through the funnel, and avoid overfilling with NaOH.
    - Avoid contact with NaOH. In the event of contact, wash the area immediately with water and notify the Instructor.
    - Dispose of chemical waste in the designated container.

Experiment 21 — *Electrical Conductivity of Aqueous Solutions*

1. See the Glossary, Appendix I.
2. (a) the light bulb glows brightly; (b) the light bulb glows dimly
3. (a) strong electrolyte; (b) weak electrolyte; (c) strong electrolyte; (d) weak electrolyte; (e) strong electrolyte; (f) weak electrolyte
4. NaCl(s) is a solid salt, and NaCl(aq) is the salt dissolved in aqueous solution.
5. (a) ionized H^+(aq) + I^-(aq); (b) nonionized HF(aq); (c) ionized Sr^{2+}(aq) and 2 OH^-(aq); (d) ionized Ag^+(aq) and NO_3^-(aq); (e) nonionized Ag_2SO_4(s)
6. The electrodes and beakers must be rinsed with distilled water to avoid a false-positive conductivity test for a weak electrolyte.
7. • Wear safety goggles; be careful to avoid contact with the chemical solutions.
    - Do not touch the exposed wire electrodes, as the wires can give a serious shock.
    - Dispose of chemical waste in the designated container.

Experiment 22 — *Activity Series of Metals*

1. See the Glossary, Appendix I.
2. (a) 0 (b) –1 (c) +7 (d) +1 (e) +5
3. (a) $Cr_2O_7^{2-} + 6\ Fe^{2+} + 14\ H^+ \rightarrow 2\ Cr^{3+} + 6\ Fe^{3+} + 7\ H_2O$
4. (b) $Cr_2O_7^{2-} + 6\ Fe^{2+} + 7\ H_2O \rightarrow 2\ Cr^{3+} + 6\ Fe^{3+} + 14\ OH^-$
5. Mg > Mn > Ni > Ag
6. • Wear safety goggles; be careful when igniting sulfur using the lab burner.
    - Avoid contact with conc H_2SO_4 and HNO_3. If contacted, wash immediately with water.
    - Avoid breathing any of the following gases, which are highly irritating and require proper ventilation: SO_2, H_2S, NO, NO_2, and NH_3.
    - Dispose of chemical waste in the designated container.

Experiment 23 — *Organic Models and Functional Groups*

1. See the Glossary, Appendix I.
2. (a) single bond (one e– pair)
   (b) double bond (two e– pairs)
   (c) triple bond (three e– pairs)
   (d) hydrogen atom
   (e) carbon atom
   (f) oxygen atom
   (g) chlorine atom
   (h) bromine atom
   (i) iodine atom
   (j) nitrogen atom
3. A black ball and green ball joined by one rigid connector: C—Cl.
4. A black ball and red ball joined by two rigid connectors: C = O.
5. (a) methyl; (b) ethyl; (c) isopropyl; (d) propyl
6. 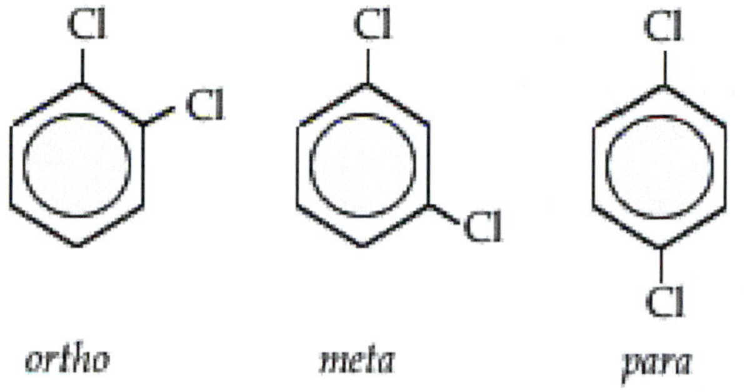

7. (a) organic halide
   (b) alcohol
   (c) phenol
   (d) ether
   (e) amine
   (f) aldehyde
   (g) ketone
   (h) carboxylic acid
   (i) ester
   (j) amide

Experiment 24 — *Separation of Food Colors and Amino Acids*

1. See the Glossary, Appendix I.
2. The "origin" is the line on a chromatogram where a drop of sample is initially spotted.
3. The "solvent front" refers to the leading edge of the solvent on a chromatogram.
4. The process of "spotting" is putting a small drop of sample on the paper chromatogram; a glass capillary tube is used to "spot" the chromatogram.
5. A "lane" is the path of a component in a sample as the solvent travels up the chromatogram. By convention, the lanes are numbered from left to right (e.g., lanes 1–4). The first component in a given lane has traveled the farthest and is closest to the solvent front.
6. Wear latex gloves when handling the chromatogram to avoid contamination.
7. The R_f value of the amino acid is 0.33.
8. A pen should not be used on a paper chromatogram because ink can dissolve in the solvent and confuse the results.
9. • Wear safety goggles; handle capillary tubes carefully to avoid cuts.
   • Do not use the organic solvent near a laboratory burner flame.
   • Avoid splashing when adding the solvent into the developing chamber.
   • Practice spotting on chromatography paper to avoid making large spots.
   • Avoid breathing and touching ninhydrin while spraying the chromatogram.
   • Dispose of chemical waste in the designated container.

Experiment 25 — *Laboratory Instruments and Techniques*

1. See the Glossary, Appendix I.
2. (a) pp. 1–2
   (b) pp. 3–5
   (c) p. 6
   (d) p. 18
   (e) p. 18
   (f) p. 19
   (g) p. 19
   (h) p. 22
   (i) p. 30
   (j) p. 30
   (k) p. 46
   (l) p. 56
   (m) p. 58
   (n) p. 71
   (o) pp. 80, 90
   (p) pp. 91, 102
   (q) p. 139
   (r) p. 149
   (s) p. 159
   (t) p. 169
   (u) p. 180
   (v) p. 192
   (w) pp. 202–4
   (x) p. 222
   (y) p. 232
   (z) p. 308
   (aa) p. 309
   (bb) p. 310
   (cc) p. 311
   (dd) p. 312

3. • Read the decigram, centigram, and milligram balances without touching the balance.
   • Read the thermometer without touching the tip of the thermometer.
   • Wear safety goggles when performing a flame test.